THE PERSONAL COMPUTER
Operating, Troubleshooting, and Upgrading

Second Edition

Mike Mutasem Awwad
DeVry Institute of Technology

Upper Saddle River, New Jersey   Columbus, Ohio

Library of Congress Cataloging in Publication Data

Awwad, Mike Mutasem.
    The personal computer : operating, troubleshooting, and upgrading / Mike Mutasem Awwad.
    --2nd ed.
        p. cm.
    ISBN 0-13-020039-5
        1. Microcomputers. I. Title
QA76.5. A 94 2001
004.165--dc21   00-036689

Vice President and Publisher: Dave Garza
Editor in Chief: Stephen Helba
Assistant Vice President and Publisher: Charles E. Stewart, Jr.
Production Editor: Alexandrina Benedicto Wolf
Design Coordinator: Robin G. Chukes
Cover Designer: Linda Fares
Cover Image: Photonica
Production Manager: Matthew Ottenweller
Marketing Manager: Barbara Rose

This book was set in Times Roman by Mike Awwad. It was printed and bound by Banta Book Group. The cover was printed by Phoenix Color Corp.

10 9 8 7 6 5 4 3 2 1
ISBN 0-13-020039-5

Preface

This book is designed to teach beginning students the fundamental skills required to operate, troubleshoot, and upgrade minicomputer systems. This book is also designed to provide insight and information on various components of a PC. This valuable information can be an aid to decision making and should lead to more effective choices of computer equipment. No prior knowledge of computer or electronics theory is required. Instead, this book is designed to develop the essential skills needed to operate and upgrade a personal PC. Emphasis is placed on the practical hands-on experience students require to become a professional computer technician.

The second edition of this book is designed to prepare the students from the introductory concepts to the analysis and services of a complete computer system. It can also be used for high school electronics programs and industry-based training programs, such as A+ certification, or as a reference for practicing technicians. The text is organized into seven chapters:

- Chapter 1: Computer Fundamentals—It presents to the student the history, fundamental concepts, and basic operating system architectures of computers.
- Chapter 2: Configuring and Upgrading PC Hardware—It teaches the techniques and skills needed to upgrade and configure a personal computer hardware.
- Chapter 3: Windows 95/98/2000 Command Lines—It shows how to use Windows 95/98/2000 command lines needed to operate, troubleshoot, and upgrade a PC.
- Chapter 4: Configuring and Operating Windows 95/98/2000—It is designed to give students an understanding of how Windows 95 operates. It also provides practical hands-on examples to perform everyday tasks in a Windows 95/98/2000 environment.
- Chapter 5: Computer Service and Support—It describes the skills and tools necessary to troubleshoot, diagnose, and service a computer system and printers.
- Chapter 6: Computer Networks—It exposes the student to the theory of computer networks, including their services and protocols.
- Chapter 7: The Internet—It summarizes the basic concepts of the Internet. In addition, students will learn how to create web pages using HTML.

This book could be used as a first semester manual to introduce computer fundamentals. It also provides all the necessary prerequisite material needed for a computer networks course. I have provided an adequate amount of theory to explain the topics and allow for further exploration if a student so desires. The reading level has been kept as simple as possible. The text follows a consistent format and pedagogy. On the other hand, I would be happy to hear from any readers about how to improve this text.

This edition redesigned the old 6 chapters into 5 new ones and added 2 additional chapters. The new material added to create the second edition is found in chapters 1, 2, 4, and 5. Chapters 6 and 7 were added to include computer networks and the Internet. The text is designed to give students a practical and step-by-step approach to maintaining and servicing a computer system.

Mike Mutasem Awwad
DeVry Institute
630 US Highway One
North Brunswick, New Jersey 08902-3362
E-mail MAwwad@admin.nj.devry.edu

To my parents, wife, son Adam, brothers and sisters for their love and encouragement

Contents

1

Computer Fundamentals

In this chapter, you will learn the history of how computers came about, and learn and understand basic computer terminology. In addition, you will learn how Microsoft's operating system Windows 98 works.

History of the PC

A hobbyist named Ed Roberts in 1975 using Intel's CPU (Central Processing Unit) 8085 designed the first PC (Personal Computer) called ALTAIR. Around 1978, Bill Gates (a dropout of Harvard University) and his programmer partner since high school Paul Allen started a company called Microsoft. Bill Gate's college roommate Steve Ballmer joined Microsoft later. The three designed the first Microsoft computer languages such as BASIC, Assembly, and Cobol.

In 1978, from a garage, Steve Jobs and his friend Steve Wozniak designed the first personal computer, the Apple II. The Apple II was not accepted in the industry because it did not have any useful functions. It needed what is called a killer application (killer app). Every technology needs a killer app to be useful. A killer app is a useful function provided by the new technology. For example, a killer app for a telephone is when two people can talk.

Harvard business graduate school's Bob Frankston and Dan Bricklin created the first killer app for Apple II. The killer app was called VISCALC. VISCALC was spreadsheet software that made accounting calculations easier with faster speed and accuracy. Although VISCALC was not patented, it made Apple II PCs a hit.

Around 1979, IBM began to notice Apple II's dominance in the market. They had no control over it, and were going to do something about it. By August of 1979, IBM began to design their first PC using an open architecture design. An open architecture design is the designing of a PC using other manufacturer's products. After IBM had designed the PC, it needed software and an operating system to run it. They turned to Microsoft for help. Microsoft was the biggest supplier of software at the time, but had nothing to do with operating systems.

Microsoft told IBM that they were not in the business of operating systems and turned them to Gary Kildall's Digital Research. Gary Kildall is the founder of the first operating system (CP /M). For some reason, IBM did not get the operating system from Digital Research and went back to Microsoft. Bill Gates was not going to let another opportunity pass him by and told IBM that his company would provide IBM with an operating system as soon as possible.

Microsoft's Paul Allen went and hired an outside programmer Tim Paterson to design an operating system for Microsoft for $50,000. The new operating system was very similar to Digital Research's CP /M. Microsoft called it PC DOS and licensed it to IBM for $50 per PC. IBM did not own PC DOS.

On August 1981, IBM launched its first IBM PC with PC DOS installed. Again for the IBM PC to be accepted in the industry, it needed a killer app. The killer app for the IBM PC was another spreadsheet called Lotus 123.

Since IBM did not own PC DOS, other manufacturers were able to design PCs similar to IBM (IBM clones). This was achievable, because IBM used open architecture design in their PCs. In 1982 a company called COMPAQ did just that. They designed an IBM clone that was cheaper and ran PC DOS. Many companies were successful. COMPAQ made money because their PCs were relatively cheaper than IBM PCs. Intel was selling CPUs to anyone who wanted to design a PC and Microsoft was selling PC DOS for every new PC sold.

IBM was losing their big PC dominance in the market. IBM PC clones were running the same operating system and applications at much lower prices. To take out DOS, IBM needed to have its own operating system. IBM turned again to Microsoft and hired Microsoft to design the new operating system called OS/2. OS/2 was equivalent to what Windows 95 today is. It was a GUI (Graphical User Interface) operating system. As Microsoft worked on OS/2, it was designing its own GUI, Windows. Windows is a GUI application that runs on top of DOS. This was Microsoft's answer to a GUI operating system. In early 1990, Microsoft decided to drop the OS/2 project and released Windows 3.0. Five years later, Microsoft released Windows 95 (GUI operating system). Most of the ideas for Windows 95 were actually invented over 25 years ago.

In 1971 Xerox the copier company formed the PARC (Palo Alto Research Center) in Palo Alto, California to work on what they called the "paperless office". In 1979, Steve Jobs of Apple Corp. visited the PARC research facility. PARC researchers showed Steve Jobs three technologies they were working on. They were:

- GUI (Graphical User Interface)—A user uses a pointer (mouse) to point to pictures and pull down menus to execute commands. A user-friendly interface.
- Ethernet Network—Computers were able to share resources, printers, file, and e-mail. The founder of the company 3COM, Bob Metcalf, was the PARC researcher that invented Ethernet networks.
- Object Oriented Programming—This is what today's Visual programming is (i.e., Visual BASIC, Visual, C++, …etc.)

Since Apple II was losing market share to IBM PCs, they immediately hired 100 engineers to work on GUI. IBM PCs was taking the market away from Apple II because of the software that ran on IBM PCs. The new Apple Macintosh (MAC) was a GUI operating system that IBM did not have. Just like any other PC, it needed a killer app. Apple then hired Microsoft to design software for their MAC. In 1984 the MAC was launched. Even with Microsoft software, it did not succeed. It cost considerably more than an IBM PC.

The actual killer app for the MAC came from the XEROX PARC research center. Since the dot matrix printer produced prints similar to a typewriter, John Warnock (a PARC researcher and the founder of the company Adobe) created the new laser printer. Laser printers with software now print exactly what you see on a GUI screen. This was not achievable using a dot-matrix printer.

The MAC started to use the Adobe software for their laser printers and "desktop publishing" took off from there. MAC machines started to pick up pace in the market. Microsoft noticed the threat, and created Windows to compete with the GUI MAC. Apple sued Microsoft alleging that they stole the Windows ideas from their MAC. Apple lost and Windows 3.0 was released in early 1990. Windows 3.0 was easier to use than a MAC. On August 24, 1995, with a campaign of over 300 million dollars, the new version of Windows called Windows 95 was released. Windows 95 is a GUI operating system with many problems. The Windows 98 release was intended to fix many of the problems Windows 95 had. The Windows 2000 Professional release is a combination of upgraded Windows 98 and Windows NT.

Computer Basics

There are a wide variety of personal computers (PCs) in today's market. Similar hardware and software and firmware products can be found on many of the PCs.

- Hardware—It is the physical components of a PC, such as a monitor, minitower, keyboard, or a mouse.
- Software—It is a preprogrammed set of instructions that are used to make the computer do the tasks we want it to do. The software always determines what hardware your PC should have.
- Firmware—It is software that is permanently stored on a chip or a disk. Instructions are only read from firmware.

The computer consists of five major components that make up its physical architecture:

- Input—Such as, a keyboard or a mouse.
- Output—Such as, a monitor or a printer.
- RAM (Random Access Memory)—It is referred to as the read/write memory. RAM is volatile (voltage dependent). Information will be erased when the voltage is turned off. When the instructions are typed on the keyboard, they are stored in RAM. RAM transmits data at speeds of 6 to 10 MHz (megahertz). MHz means millions of cycles per second. When a computer program starts, it is loaded into RAM memory for use.
- ROM (Read Only Memory)—It contains all of the instructions your PC needs to run. ROM is non-volatile (voltage independent). This type of memory is permanent and cannot be erased.
- CPU (Central Processor Unit)—It is a small electronic circuit that acts as the "brain" of the computer. It picks up instructions from RAM, executes them, and sends them to an output. The CPU executes instructions that are stored in RAM memory only.

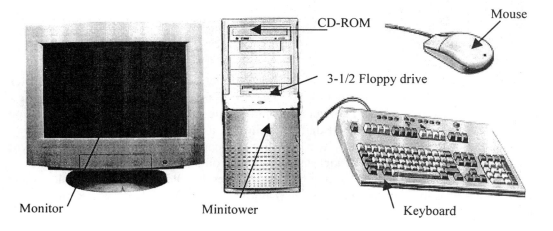

Figure 1.1 Basic hardware of a PC

The American Standard Code for Information Interchange (ASCII) requires one byte (8 bits of data) to represent each character on the keyboard. A program is broken into several sections. A computer understands only electronic signals that mean "on" or "off". These signals, which can be noted by either a 1 or a 0, are called bits. Bits are the smallest unit of information a computer can recognize. When bits are combined into a group of eight, they are called one byte. A word is a combination of two or more bytes. A field is a group of words. A record is a group of fields and a file is a group of records. A program is a group of files.

The program is a set of instructions that are stored in files to tell the CPU what task to perform. Each file on the disk must have a label. This label is called a filename. The filename comes in two parts, the filename and its extension. In DOS versions 1.xx to 6.xx the filename can be from one to eight characters in length and the extension can be zero to three characters in length. You cannot include the characters * . \ / ? or space as part of the filename or extension name. In the Windows 95/98 versions, you can have up to 255 characters for a filename including space.

SOFTWARE.EXT

Filename   Extension

Every file should have an extension to distinguish it from other files that have the same filename. Here is a short list of extensions and what they stand for: .EXE- (Executable), .BAT- (Batch), .COM-(Command), .SYS-(System), .DOC-(Document), and .TXT-(Text). You may even name a disk internally. The internal name you give a disk is called the volume label.

DOS displays a symbol called a prompt to let you know it is ready to receive a command. The prompt consists of the default drive, the symbol ">" and a flashing "_ " called a cursor. This prompt also identifies the root directory. The root directory identifies the top-level contents on a logical disk. It contains the name, size, date, and time a file was last changed. It might also contain other directories called subdirectories. A subdirectory is a table of contents on a disk or just a file whose contents happen to be the directory entries for all the files (and any subdirectories) that are in that subdirectory. The first two entries in a subdirectory are always ".", which refers to this subdirectory itself and "..", which refers to the parent directory. Even if you delete every file and subdirectory on a disk, you still have a root directory. Otherwise, you would not be able to create new files and keep track of their existence.

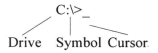

Drive   Symbol Cursor

The foundation software that controls your PC's basic functions is called the operating system. The most popular operating system in today's market is Windows 98. It is a GUI (pronounced "gooey") operating system. A GUI conveniently allows the user to choose icons (pictures) on the screen that represent commands, rather than requiring the user to type commands. In older versions of the Microsoft operating system DOS (Disk Operating System), the user must type commands when he or she wants a function to be carried out. This means memorizing the commands you use most often, or looking them up frequently. Older DOS versions are command driven, which means those earlier versions of Windows such as MS Windows 3.x must run on top of DOS. Microsoft Windows 3.x is also referred to as a GUI.

Since Microsoft Windows 3.x is placed on top of DOS and makes it easier to use, it is not an operating system, but a graphical environment. Windows 95, which was released after Windows 3.x, is a GUI operating system. Windows 98 is the latest version of a GUI operating system. One special advantage of Windows is its multitasking capabilities. Multitasking means that the operating system can run two or more programs simultaneously.

The operating system allows the use of certain devices with the computer such as displays or monitors, floppy drives, fixed drives, printers, and modems. The operating system assigns names for such devices:

Device	Operating System Device Name
Drive	Letters A to Z
Printer	LPT1, LPT2, LPT3
Display	Con (Console)
Modem Port	COM1, COM2

Table 1.1 DOS device names

Keyboard and Mouse

The 101-key enhanced keyboard is the standard keyboard used with most of today's PCs as their primary source of input. It features 12 function keys across the top of the alphanumeric keys and three lights to indicate when the NUM LOCK, CAP LOCK, and SCROLL LOCK keys are toggled "on."

- <Enter> Key—When pressed, it transmits commands to the RAM memory of the PC. You must press the <Enter> key to tell DOS to execute a command.
- <Ctrl> Key—When pressed, the (Control) key is used with other keys to do a specific task.
- <Alt> Key—When pressed, the (Alternate) key is used with other keys to do a specific task.
- <Shift> Key—When pressed, it is used to capitalize letters or to obtain characters or symbols on the top half of some keys.
- <Backspace> Key—When pressed, it erases a character or a space to the immediate left of the cursor.
- <Caps Lock> Key—When locked, the (Capital Lock) key makes all the alphabetical characters appear in upper case.
- <Num Lock> Key—When locked, the (Numerical Lock) key determines whether the numeric keypad located on the right-hand side of the keyboard will function as a ten-key calculator pad or as cursor movement keys.
- <Print Screen> Key—When pressed, it sends the content on the screen to a printer.
- <Esc> Key—When pressed, the (Escape) key cancels a line you have typed.
- <Insert> Key—When pressed, it inserts one or more characters.
- <Delete> Key—When pressed, it deletes a character in front of the cursor.

Another type of an input device is the mouse. The mouse is a hand-held device used to control the position of a pointing tool. Use the mouse as an extension of your hand. Place the palm of your hand on the top of the mouse, with the cable extending forward from your fingers. Place your thumb on the left side, rest your index and middle fingers lightly on the buttons and clasp the right side of the mouse with the remaining fingers. Moving the mouse requires a smooth, clean surface and approximately a square foot of space. That is why mouse pads are commonly used.

DOS Architecture

There have been six versions of MS-DOS introduced over approximately 12 years (1981 to 1993), namely versions 1.xx, 2.xx, 3.xx, 4.xx, 5.xx, and 6.xx. The xx is used to indicate a incremental minor enhancement to the previous version. Windows 95 contains DOS 7 with Windows 4.0 combined. Table 1.2 shows the features of the six versions of DOS and how they have evolved, version to version, over time:

DOS Version	Features
DOS 1.xx	• Introduced in 1981 with no provisions for networking. • DOS 1.1 supported double-sided 320-KB floppy drives. • DOS 1.25 designed for non-IBM hardware (clones).
DOS 2.xx	• Introduced in 1983 for IBM PC XT. • Supported 10-MB hard drive, serial interfaces, and an additional 3 expansion slots. • Introduced the hierarchical "tree" structure to the DOS file system.
DOS 3.xx	• Introduced in 1984. • Supported hard drives larger than 10-MB and enhanced graphics formats. • DOS 3.1 was the first DOS to support networking. • DOS 3.2 introduced the XCOPY command, supported IBM's Token Ring network topology, and allowed for 720-KB 3 ½ inch floppies. • DOS 3.3, introduced in 1987. supported 1.44-MB floppy disks and logical partition sizes of up to 32 MB.
DOS 4.xx	• Introduced in 1988, with the first graphical DOS shell. • Ability to use the mouse in DOS shell.
DOS 5.xx	• Introduced in 1991 with ability to load drivers in upper memory. • EDIT.COM utility and QBASIC.EXE program were introduced. • Commands such as DOSKEY.COM, UNFORMAT.EXE, and UNDELETE.EXE were introduced.
DOS 6.xx	• Introduced in 1993with new utilities such as new antivirus and backup software, a defragmentation utility, drive compression, and the ability to pool EMS (Expanded memory) and XMS (Extended memory) using EMM386.EXE.

Table 1.2 Six versions of DOS

Booting Process

The main function of an operating system is to be an interface between the software, hardware, and the user. For example, the most famous operating system MS-DOS (Microsoft Disk Operating System) controls the way the computer uses application programs, text files, and games. DOS allows the user to create files, use application programs, and manage files on the computer. It is also responsible for writing and reading from a floppy and a hard drive. The term cold boot refers to turning the power switch on. The term warm boot refers to while the computer is on, pressing the keys <Ctrl>, <Alt> and <Delete> at the same time, clearing the RAM and reloading all the system files. When the power is first turned on, the ROM chips on the motherboard start to execute of the following four steps in sequence automatically:

1) The POST (Power-On-Self-Test) is performed. The POST is a series of diagnostic programs designed to check the basic system components.

   a) Basic System Test—Checks the operation of the CPU, system bus, and the memory segment containing the POST ROM.
   b) Extended System Test—Checks the system timer and ROM BASIC interpreter.
   c) Display Test—Checks the hardware that operates the video signals. It will test the video RAM for a default display adapter and the video signals that drive the display.
   d) Memory Test—Checks the computer's memory using the writing and reading back of patterns.
   e) Keyboard Test—Checks the keyboard interface and looks for stuck keys.
   f) Cassette Test—Checks the cassette recorder interface of the PC.
   g) Disk Drive Test—Checks to see what disk drives are installed. If there is a disk installed, then the disk interface card is tested.
   h) Adapter Card Test—For PS/2 systems this test checks and configures an installed adapter card.

Generally, when the POST encounters an error on an IBM PC, it will indicate the type of error by a POST error code (number) or an audio sound. The POST error codes are numbers whose values indicate the type of problem encountered. Using the system speaker, The POST can also indicate the type of problem by a series of short beeps. Chapter 5 covers the POST error codes and audio sounds when a problem occurs.

2) DOS then loads the bootstrap program. The bootstrap program will look to see if the operating system files (DOS) are located in floppy drive A. If they are not present, it will look for them on the hard drive C. When the operating system files are found, the bootstrap program then initiates the booting process. The booting process is the process of transferring the DOS kernel files (IO.SYS, MSDOS.SYS) and COMMAND.COM to RAM. Once all three files reside in RAM, they form what most users think of as DOS.

   a) IO.SYS file—It is the first DOS kernel sometimes referred to as BIOS (Basic-Input-Output-System). It is a group of software routines that controls the computer's hardware. These routines communicate with the BIOS chip on the motherboard to perform such functions as moving drive heads or printing characters to the monitor. To prevent the user from accidentally deleting IO.SYS, the file is hidden.

   b) MSDOS.SYS file—It is the second DOS kernel sometimes referred to as BDOS (Basic Disk Operating System). From the name (Disk Operating System), it is a software routine that performs the disk filing (disk I/O tasks). It is also responsible for performing high-level control of information flow between the various parts of the computer. To prevent the user from accidentally deleting MSDOS.SYS, this file is also hidden.

   c) COMMAND.COM file—It is sometimes called the command processor, command interpreter, or DOS shell. It provides the command line interface that the DOS user sees. It is responsible for displaying names and sizes of files on the disk. It is also responsible for displaying the content of files and determining the proper action when a problem occurs. All the internal commands of DOS are located in COMMAND.COM. The internal DOS commands are the basic DOS commands; they include DIR, COPY, DEL, CLS, etc.

Remember that the internal DOS commands are a part of COMMAND.COM. They always reside in memory; thus can be executed at any time at any prompt. On the other hand, the external DOS commands are located on a disk, usually in a DOS directory. The external DOS commands are only loaded into memory whenever they are needed. COMMAND.COM processes (interprets) the external commands as they occur. The external DOS commands have either a .COM or an .EXE extension.

3) DOS will search for a file called CONFIG.SYS before it loads COMMAND.COM. If found, the commands inside it are executed (transferred) to memory. This file contains special commands used to configure (set up) your DOS for use with devices or applications. CONFIG.SYS file will be discussed in detail later in this chapter.

4) DOS will finally search for a file called AUTOEXEC.BAT. This file can perform any commands normally executed at the prompt. The AUTOEXEC.BAT file will also be discussed in more detail later in this chapter.

The Memory Map

The CPU memory is simply arranged as a block where any byte of information is directly available by calling its address. The address lines of the 8088 microprocessor connect to some circuitry called address decoder logic which translates the address numbers requested by the microprocessor to the matrix arrangement of the memory chip. RAM (Random Access Memory) is the working memory of the computer. For example, when you run a program, it is loaded from your hard drive into RAM and remains there while you use it. The work on a disk must be saved, because whatever is in RAM will get erased when the computer is turned off. RAM chips are the physical components that contain the memory. They are grouped in rows called banks.

Most RAM chips are made of capacitors. A capacitor is an electronic device that holds an electric charge. However, the insulators inside the capacitor are not perfect. The capacitor must leak charge for a few milliseconds to be useful. In the PC world, there are some special circuits that are used to periodically recharge the capacitors and refresh memory. Those circuits are called dynamic memory (DRAM). Dynamic RAM allows current to flow and continue on its way.

The other types of RAM chips are SRAMs. SRAM (Static RAM) tries to trap electricity and hold it in place. The kind of memory a computer can only get information from (read) and not store information in (write) is called a ROM (read-only-memory). A ROM chip has access time between 50 and 200 billionths of a second. It retains its content when a computer is turned off. This type of memory typically holds the start-up programs that prepare the computer for use.

There are two major areas of computer memory; primary and secondary. Primary memory is a memory that is immediately accessible to the CPU. It is directly addressable with address lines or through the I/O ports of a computer. It has limited capacity and is used for short-term storage. Secondary memory on the other hand is a long-term storage area and is not immediately accessible to the CPU. Any information that needs to be used from the secondary memory must be transferred to the primary memory.

When IBM released its first personal computer in 1981, it came up with an 8088 microprocessor that supported 20 address lines. Having 20 address lines gave the CPU the ability to access 1 MB ($2^{20} = 1,048,576$ bytes = 1 MB) of memory. The designers of DOS split this memory range into two sections. The first 640 KB was called the conventional memory and the rest (384 KB) was called the upper memory.

Upper Memory

Sometimes called reserved memory, it is basically reserved for video display and other hardware devices. When your computer displays letters on the screen, the letters are first placed in the video memory. The video memory, however, only consumes part of the upper memory area, leaving part of the memory available for your programs to use. Normally, DOS loads device drivers and memory-resident programs into the 640-KB program space. By moving these programs to the upper memory area, you free up more of the 640-KB program space for use by your programs. A memory-resident program is a program that remains in your computer's memory after you run it.

Conventional Memory

The conventional memory is used to execute programs. It is where DOS loads and runs programs. Lower conventional memory is reserved for use by the computer, but from about the 2-KB mark on up to 640 KB, you can run applications. This doesn't mean that the full 640 KB (or 638 KB) is available to execute programs. DOS may require more conventional memory to execute big executable files.

For example, to run WordPerfect 5.1 for DOS, the executable file WP.EXE must be loaded to the conventional memory. When the computer starts, it loads DOS (the operating system) into the conventional memory. Every PC, even a Pentium, uses the conventional memory, because previous versions of DOS treated the microprocessor like a fast 8088 with 1 MB of memory. No matter how much memory the computer has, DOS-based programs always run within the conventional memory. If a program cannot fit within the 640-KB memory region, DOS cannot run the program. However, the programs can store their data in extended and expanded memory, making more room in the conventional memory for program instructions. When users perform memory management within DOS, their goal is to free up as much of the conventional memory as possible for use by their programs.

Today's programs have grown in size to the point where many programs cannot hold their instructions and data in the 640-KB program area. Beginning with the 286 processor, the PC can use a second memory type called the extended memory.

Extended Memory

The extended memory is linear and continues where DOS 1 MB leaves off. Depending on the model, PCs can hold up to 128 MB of extended memory. The MEM command displays the system's memory use, the amount of extended memory the system contains, and how big the conventional memory is. When a PC is purchased, the PC normally comes with 8, 16, 32, or up to 256 MB of memory. The first 1 MB is used for the conventional and upper memory. More extended memory can be added to the PC by purchasing chips called SIMMs (Single Inline Memory Modules). SIMM chips are installed on your system's motherboard. The more extended memory you have in your system, the faster the PC operation will be.

In order to use extended memory, a program must switch the microprocessor to protected mode and then switch back to real mode before quitting. When a software program runs in real mode it runs in conventional memory using DOS-compatible software. This DOS-compatible software emulates 80286 or higher systems as if they were 8088 systems. Software is said to be running in a protected mode if it uses the extended memory to execute programs.

An example on the use of the extended memory is the loading of Windows 3.1. Windows 3.1 requires a minimum of 2 MB of extended memory. Let's assume that the PC has only 2 MB of memory on board. When the WIN.COM file is loaded in the conventional memory, there are instructions in it to transfer more than 2 MB of files to the extended memory. Only 2 MB of files will be transferred to the extended memory. The rest are placed on the hard drive in an area called virtual memory. Whenever the CPU needs files from virtual memory, it swaps with the extended memory files, slowing down the operation of the PC. Therefore, the more extended memory the PC has, the less swapping with virtual memory will be, resulting in faster execution of the programs. Some software will require a minimum amount of extended memory, which means that it needs to transfer at least that size of files to the extended memory for the application to start.

Remember that when the system boots up, the DOS operating system (IO.SYS, MSDOS.SYS, and COMMAND.COM) is transferred to the conventional memory. DOS can be removed from the conventional memory to the first block of the extended memory. Freeing up the conventional memory will provide enough space to execute large executable programs. For example, if the MEM command was executed and the statement "The largest executable program size is 450 KB", appears on the screen, then the conventional memory can only execute files that are 450 KB or less in size. Moving files from the conventional memory to the extended memory will increase the conventional memory area; thus programs larger than 450 KB can be executed.

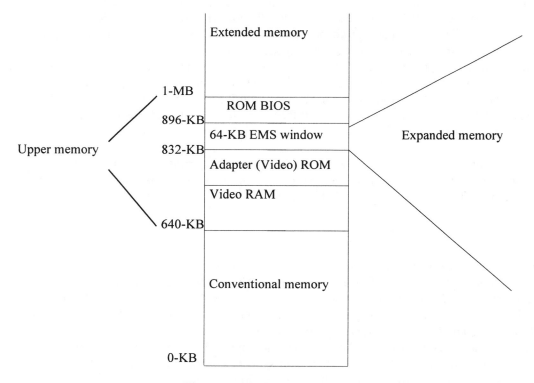

Figure 1.2 DOS memory map

Expanded Memory

Expanded memory is mostly used to free up conventional memory. EMM (Expanded Memory Manager) is software that tricks DOS into thinking that memory outside the 1-MB limit is located at addresses within the 1-MB limit. Before the upper memory area can be used, DOS must be informed to support it. The expanded memory cannot be accessed directly as extended memory can. It uses a 64-KB memory segment called a window. The window is usually installed as a segment in the first 1 MB somewhere above the video RAM area at a place that is usually not used otherwise. The window looks out to a special memory adapter card that contains a number of 64-KB segments. The card will have circuits that allow it to access a 64-KB segment in the first 1 MB. Then as the 64-KB segment fills up with data, that segment is switched back to the memory card and a fresh 64-KB segment is switched into the 1-MB memory area at the same address as the one that was switched out. This segment switching can go on and on as long as there are empty 64-KB segments on the memory card.

CONFIG.SYS File

Each time the system starts, DOS searches the disk's root directory for a special file named CONFIG.SYS. If DOS locates the file, DOS uses the entries the file contains to configure the system. If the file is not found, DOS uses its default settings. The CONFIG.SYS file is similar to the IO.SYS file. The IO.SYS kernel file loads certain built-in device drivers (software routines), such as the console device driver and mouse driver to operate your system.

CONFIG.SYS extends the use of additional hardware and software by loading their respective drivers. When different hardware components are installed, there may be times when there is a need to load special software, called device drivers, into the computer's memory. A device driver is special software that lets your operating system (DOS) or (Windows) recognize and use the hardware device. To load the device driver, an entry is normally placed in the CONFIG.SYS file. Devices are stored in files on disk, much like programs. Most drivers use the .SYS extension, such as MOUSE.SYS. Here is a typical CONFIG.SYS file:

```
DEVICEHIGH=C:\DOS\ANSI.SYS
DEVICE=C:\DOS\HIMEM.SYS
DEVICE=C:\DOS\EMM386.EXE NOEMS
DOS=HIGH, UMB
BUFFERS=20
FILES=30
BREAK=ON
```

The entry DEVICEHIGH=C:\DOS\ANSI.SYS is used to install a device driver into the upper memory area. It loads the device driver ANSI.SYS file into the upper memory. If DOS is unable to fit the device driver into the available upper memory, DOS will load the driver into the 640-KB program area. ANSI.SYS is a device driver that gives DOS additional control of the monitor screen and keyboard devices that are beyond the control built into the operating system.

The DEVICE entry installs a device driver. In the first line in the above CONFIG.SYS file, DEVICE=C:\DOS\HIMEM.SYS loads the device driver HIMEM.SYS from the DOS directory to memory. The HIMEM.SYS device driver is used to manage the extended memory. It makes the upper and the extended memory available for use. The file HIMEM.SYS as a file allows access to the extended memory. The DOS=HIGH command will move the operating system DOS from the conventional memory.

The EMM386.EXE RAM is required to make the upper memory available for the creation of UMBs (Upper Memory Blocks) into which you can load TSRs (memory-resident programs) and device drivers. The DOS=HIGH, UMB entry loads the operating system from the conventional memory to the first 64 KB of extended memory. This will free up more conventional memory. The UMB option is necessary to create UMBs into which TSRs or device drivers can be placed to save space in DOS 640 KB conventional memory. The EMM386.EXE device driver allocates extended memory for use as expanded memory. In this case, the parameter NOEMS tells DOS that expanded memory will not be used. Instead, it is simply using the device driver to provide support for the upper memory area. The upper memory can be used with the commands LOADHIGH in AUTOEXEC.BAT and DEVICEHIGH in CONFIG.SYS.

The BUFFERS entry sets the number of file buffers DOS needs to reserve for more efficient transfer of data to and from a disk. A buffer is approximately 512 bytes in size. Since data is stored in sectors of 512 bytes, the data is accumulated in the buffer until it is full, then written to a disk, thus reducing disk activity. The FILES entry sets the number of files that can be open at one time. If no FILES statement is used, the minimum number of files is used. This is not enough for today's popular software applications. The BREAK=ON entry informs DOS to monitor the keys <Ctrl> and <C> to see if they are pressed during a DOS command execution. If the keys are pressed, then the DOS command will be interrupted and halted.

You may edit the CONFIG.SYS file using the EDIT command. Every time changes need to be made to CONFIG.SYS file, the PC must be rebooted for the changes to take effect. The drives and paths must be included in CONFIG.SYS, because AUTOEXEC.BAT, which contains path statements CONFIG.SYS might use, is executed after CONFIG.SYS.

AUTOEXEC.BAT File

After the system files and CONFIG.SYS file are transferred to memory, DOS searches the disk's root directory for a special batch file named AUTOEXEC.BAT. AUTOEXEC.BAT contains a list of commands that automatically (AUTO) execute (EXEC). Commands normally found in AUTOEXEC.BAT include the PROMPT command, which defines the system prompt; the PATH command, which tells DOS where to locate the program files (within which directories) and device driver entries. Here is a typical AUTOEXEC.BAT file:

```
DATE
TIME
ECHO  Good Morning Class
ECHO OFF
DOSKEY
PATH=C:\DOS;C:\WINDOWS
PROMPT=$P$G
SET TEMP=C:\WINDOWS\TEMP
```

In the above AUTOEXEC.BAT file, the internal DOS commands DATE & TIME are executed first. The ECHO *message* entry displays a message on the screen. The ECHO OFF entry suppresses the display of all batch file commands that follow. To prevent ECHO OFF from displaying, type an @ in front of the entry. The DOSKEY entry repeats and edits DOS commands. Press the up arrow key until the previously issued DOS command is found.

The PATH entry sets a search path for DOS to find programs (i.e., commands or executable files) that are needed to run. If the program (command) needed to run is not located in the current directory, the PATH entry tells DOS in which directories to search for this program. Without a path statement, DOS looks only in the current directory for the program (command). Notice the semicolon (;) between the paths. DOS searches the current directory first, then the directories listed in the PATH statement, from left to right.

The maximum length of the PATH statement is 127 characters. If program files in a directory have the same filename but have different extensions, DOS runs them in the following order of precedence: .COM files, .EXE files, and then .BAT files. When controlling which program must run DOS will look for the typed extension after the filename. The PROMPT command allows control of how the DOS system prompt appears. PROMPT=PG will display the default drive, and will display the path (location) of where the user is located.

In DOS 6 or later the AUTOEXEC.BAT or CONFIG.SYS files may be prevented from executing automatically by pressing the key <F5> within two seconds after the message "Starting MS-DOS..." is displayed. DOS will then start a minimal system, and will not have a complete command path defined. Likewise, a user cannot open more than three files at any given time. Bypassing the system startup in this way provides the user with an opportunity to correct errors within CONFIG.SYS or AUTOEXEC.BAT. After the changes are made, the system must be rebooted using the keys <Ctrl> <Alt> and <Delete> in combination. In DOS 6 or later, to process only specific CONFIG.SYS and AUTOEXEC.BAT entries, the function key <F8> must be pressed as the system starts. DOS, in turn, will display a prompt for each CONFIG.SYS and AUTOEXEC.BAT entry. To process each entry, press Y. To bypass the entry, press N.

When Windows 3.1 is launched, it places temporary files in a directory on the hard drive called TEMP. These files start with the character ~ and have the extension .TMP. Windows swaps these files with the extended memory whenever it needs them. Windows then will delete all the files in the directory TEMP upon closing it. If the power is turned off without closing Windows, the TEMP files will not be erased. These files take up space on your hard drive and need to be removed, because they might overload your hard drive without your knowing it. They will prevent other applications from being launched because of an inadequate amount of virtual memory.

It is best to create the subdirectory TEMP in the directory Windows and place the statement SET TEMP=C:\WINDOWS\TEMP in the AUTOEXEC.BAT file. The SET TEMP command tells DOS where to place the temporary files; it does not create the TEMP directory. If there is no TEMP directory, the temporary files will be scattered all over the hard drive. Remember that anytime a command in CONFIG.SYS or AUTOEXEC.BAT is placed, the system must be rebooted for it to take effect.

Windows 3.x Architecture

Windows was developed for the 286 and 386 IBM-compatible machines. Intel's 286 and 386 processors operated in three different modes. Therefore, applications needed to run under these three different modes.

- Real Mode—This mode emulates the 8088 CPU processor. Any processor, including Pentiums, running real mode has the same limitation that an 8088 has. Real mode uses the first 20 address lines, thus it is limited to 1 MB of RAM space.
- Protected Mode—This mode was designed to make use of the 286 processor's 24 address lines. Therefore, any CPU running in protected mode has access to 16 MB of RAM.
- Enhanced Mode—This mode allowed another feature called virtual mode. Virtual mode allowed the CPU to pretend that it is a multiple of 8086 or 8088 CPUs. Each version of the multiple CPUs is called a virtual CPU and has the same restriction as a real 8088 CPU. Each virtual CPU can run separate copies of programs.

There have been seven versions of MS Windows introduced, namely, Windows 1.xx, Windows 2.xx, Windows 3.xx, Windows 95, Windows 98, Windows NT, and Windows 2000. Table 1.3 shows the features of the seven versions of Windows and how they evolved, version to version over time:

Windows Version	Features
Windows 1.xx	• Introduced in 1985 with tiling windows, mouse support, and menu systems. • Introduced cooperative multitasking. • Did not use icons.
Windows 2.xx	• Introduced in 1987 with icons added • Allowed application Windows to overlap each other as well as tile. • Support for PIFs (Program Information Files), which allowed user to configure Windows to run their DOS applications more efficiently.
Windows 3.xx	• Introduced in 1990. It allowed access to more memory than the 640-KB limit imposed by DOS. • File manager and Program Manager were introduced for network support. • Use of the 386 Enhanced Mode was introduced to allow the use of parts of the hard drive as "virtual memory." • Windows 3.1 introduced in 1992 supported only 16-bit applications, which provided better graphical displays and multimedia support. • Windows 3.1 introduced the OLE (Object Linking and Embedding) to let applications work together more easily. • Windows 3.11 (Windows for Workgroups) supported both 16-bit and 32-bit applications.
Windows 95	• Introduced in 1995. The first GUI operating system with plug-and-Play features. • Supports both 32-bit, and 16-bit drivers as well as DOS (8-bits) drivers.
Windows 98	• Introduced in 1998 with changes in the architecture of Windows 95 to make it more stable.
Windows NT (New Technology)	• Supports huge drive sizes and can access up to 4 GB of RAM. • Used for users who use large files or complex programs. • Better security and more stable than the other versions.
Windows 2000 Professional	• Introduced in February 2000. The combination of the upgraded version of Windows 98 and Windows NT4.0.

Table 1.3 Seven versions of Windows

Windows is a GUI interface that makes interfacing with a computer easier. Remember that Windows 3.x is not an operating system in and of itself, but an application that runs on top of DOS. Windows 3.x used what is called OLE. OLE (Object Linking and Embedding) is the idea that an object is not a screen representation; rather it is a document or a file. For example, one document might be linked to another. Instead of having two copies of the same data, one copy can be referred to by another document. Changes made to one document will appear automatically in the other document, which has a point embedded in it that locates the original data. In Windows 3.x, the most important system files are WIN.COM, kernel files, main system drivers, and the initialization (INI) files.

- WIN.COM—This file is used to load the startup logo screen and the kernel files and other core components.
- Kernel Files—The kernel files KRNL286.EXE and KRNL386.EXE are responsible for loading and running applications and management of resources. They control memory management and scheduling resources, which control the flow of tasks in a multitasking environment. The KRNL286.EXE file is loaded for standard (real) mode operation and the KRNL386.EXE file is loaded for virtual (enhanced) mode operation.
- Main System Drivers—the main system drivers are GDI.EXE and USER.EXE. The GDI.EXE (Graphical Device Interface) file is used to control and create the GUI. The USER.EXE file is the Windows input and output manager.
- Initialization (INI) files. The main two INI files Windows use are WIN.INI and SYSTEM.INI. The WIN.INI file contains the settings that let users personalize their Windows installations. Other programs may use WIN.INI for their settings. Some hardware settings may also be found in WIN.INI file, such as the serial and printer port settings. The SYSTEM.INI file contains the primary hardware configuration settings.

You may start Windows 3.x in two ways:

1) Standard Mode—When Windows 3.x starts in standard (real) mode, WIN.COM first loads a file called DOSX.EXE in order to provide extended memory support. DOSX.EXE then loads the standard-mode kernel, KRNL286.EXE. The KRNL286.EXE files then load Windows drivers (files with .DRV extensions), GDI.EXE, USER.EXE, and various supporting files, such as network access files. Finally KRNL286 loads the shell specified in the SYSTEM INI file, usually Program Manager.
2) Enhanced Mode—When Windows 3.x starts in enhanced (virtual) mode, WIN.COM first loads the WIN386.EXE file. The WIN386.EXE file then loads the VMM (Virtual Memory Manager) and the virtual device drivers, which are designated by the extension .386 in SYSTEM.INI files. The WIN386.EXE file then loads the kernel file KRNL386.EXE. The KRNL386.EXE file then loads Windows drivers (files with .DRV extensions), GDI.EXE, USER.EXE, and various supporting files, such as network access files. Finally KRNL386 loads the shell specified in the SYSTEM.INI file, usually Program Manager.

Windows 95/98 Architecture

Old DOS had many failings and imposed memory limitations. Windows 3.x improved the imposed memory limitations and allowed access to larger memory spaces. In addition, Windows 3.x retained some of the old DOS problems, such as UAEs (Unexplained Application Errors). Windows also had its own problems such as GPF (General Protection Fault) errors. A GPF happens when a program attempts to access part of the memory space that is reserved by another program. That is why most of the time Windows 3.x runs in the extended memory under the protection mode of the CPU.

Windows 95 was designed to solve many of the old DOS and Windows problems. It was a new operating system that worked with most of the old applications. In general, it is designed to do the same job as Windows 3.x, only better. Windows 95 is an operating system that provides an enhanced GUI. It provides an enhanced and more stable networking features than Windows 3.x, allows long filenames, offers large drive partition sizes, and offers the famous Plug-and-Play feature.

Windows 95 GUI still uses overlapping windows. Now the icon appears on what's called a desktop. An icon can either be a folder or a shortcut to an application executable file. A folder is a graphical representation of an organizational object.

The old Windows 3.x featured the cooperative multitasking model. In the cooperative multitasking model, programs that are opened simultaneously can interfere with each other causing UAEs and GPFs. To minimize UAEs and GPFs, Windows 95 uses the preemptive multitasking mode. In preemptive multitasking mode, the operating system will always control the CPU, not the programs. The operating system (Windows 95) tells the CPU to do other tasks even while other applications are running.

Windows 95/98 offers long file names with a limit of 255 characters and the file name may have spaces and some special characters. In addition, in the new version Windows 98, partitioning a hard drive can be performed to a have a drive space of up to 2,048 GB through the use of FAT16 or 2 terabytes through the use of FAT32.

Windows 95/98 uses what is called a flat memory, instead of a segmented memory model. In the flat memory model, all memory addresses have one complete segment, unlike the segmented where you have the segment address and the offset address. In the flat memory model, very large applications can be written and secured addressing can be accomplished.

Windows NT/2000 Architecture

Windows NT is a modular operating system. It is not a single large program, but instead it consists of many small software elements (modules) that cooperate to provide the system's networking and computing capabilities. Windows NT operates in one of two different modes.

- User Mode—In this mode only moderate access to Windows NT resources is granted. Every user interaction with the Windows NT environment occurs through the user mode process. If a user mode needs objects or services, the user mode must go through the kernel mode to obtain access. User mode subsystems enable Windows NT to emulate Win16-bit DOS environments, Win32-bit Windows APIs, and even OS/2 character mode.

  The user mode security subsystem is the only subsystem responsible for the logon process. The security subsystem works directly with the kernel mode to authenticate (verify) the logon process.

- Kernel Mode—This mode defines the inner operation of Windows NT. Every component in the kernel mode has priority over user mode subsystems and processes. The kernel mode blocks the access of user applications from directly interacting with NT hardware and core system services. That is why user applications must request access from the kernel mode whenever they need access to NT hardware or low-level resources.

Partitioning a hard drive in Windows NT can be performed to a have a drive space of up to 2 terabytes through the use of NTFS. NTFS (NT File System) is a 32-bit format scheme and provides better partition security than FAT32. NTFS allows file-by-file compression and no other operating system can access its partitions.

Windows NT can manage up to 4 GB of RAM (2 GB for user mode and 2 GB for kernel mode). Windows NT uses a flat 32-bit memory model. It is a model that is flat and grows based on the demand for memory. This is as opposed to models in which every section of memory has a fixed role such as in the conventional, expanded, and extended memory architecture present in DOS and Windows 3.x. Windows NT memory uses a flat 32-bit memory model that is based on a virtual memory, demand-paging method to each 32-bit and non-32-bit application. A page is a 4-KB unit of memory that the VMM (Virtual Memory Manager) program manipulates.

Both Windows 95/98 and Windows NT offer the benefits of a 32-bit operating system. Windows NT has an advantage over Windows 95/98 with its improved expanded memory management capabilities, support for preemptive multitasking and increased robustness and stability. Windows NT operating system requires more RAM that its counterpart Windows 95/98. The hardware requirements for a Windows NT 4.0 is:

486 processor, 33 MHz or better
A minimum of 12 MB of RAM (16 MB recommended)
A minimum of 110 MB hard drive

The combination of Windows 98 and Windows NT into one operating system called Windows 2000 Professional provided the benefits of both operating systems. Windows 2000 Professional, which was released on February 2000, combined the best features of Windows 98—Plug-and-Play, easy-to-use user interface, and power management—and made them better. The strength of Windows NT was integrated to provide standard-based security, manageability, and reliability.

Windows 2000 Professional is designed to be very reliable. It introduces Windows File Protection to protect core system files from being overwritten by application installs. In addition, it provides driver certification, full 32-bit operating system, and Microsoft Installer, and dramatically reduces reboot scenarios.

Windows 2000 Professional is easy to use and maintain. It provides features such as System Preparation Tool (SysPrep) to help administrators clone computer configurations, systems, and applications, resulting in simpler, faster, and more cost-effective deployment. It also provides other features such as Setup Manager, remote OS installation, multilingual support, faster performance, faster multitasking, scalable memory and processor support (supports up to 4 gigabytes and up to two symmetric multiprocessors), peer-to-peer support for Windows 9x and Windows NT, Microsoft Windows Services for UNIX 2.0, personalized menus, troubleshooters, Preview windows for multimedia, more wizards, Windows NT security model, Encrypting File System (EFS), IP security (IPSec) support, recovery console, and safe mode startup options.

Finally, Windows 2000 Professional was built for mobile users and it is Internet ready. Mobile users have features such as Hibernate. You may turn off your computer and monitor after a predetermined time, while retaining your desktop on disk. When you re-activate your computer, hibernate mode restores your programs and settings exactly as you left them. Other features for mobile users include Offline Files and Folders, Offline viewing, Synchronizing Manager, Smart Battery, Hot Docking, easier remote configuration wizards, NetMeeting, and Universal Serial Bus (USB).

Windows 2000 Professional integrated Internet Explorer 5.01 and your desktop environment with the web. Other Internet features of Windows 2000 Professional include strong development platform (support for Dynamic HTML Behaviors and XML), Search bar, History bar, Favorites, Internet Explore Administration Kit (IEAK), Auto complete, AutoCorrect, Automated Proxy, and Internet Connection Sharing. The hardware requirements for Windows 2000 Professional are:

Pentium processor, 133 MHz or better
A minimum of 16 MB of RAM (for Operating System to load only). 64 MB recommended
A minimum of 2 GB hard drive with a minimum of 650 MB of free space
Supports single or dual CPU system

Questions

1) Explain the difference between hardware, software, and firmware.
2) What are the five major components that make up the physical architecture of a minicomputer?
3) What do the following stand for?

a)	PC	j)	BIOS
b)	DOS	k)	EXE
c)	RAM	l)	DOC
d)	ROM	m)	COM
e)	MHz	n)	SYS
f)	CPU	o)	TXT
g)	ASCII	p)	BAT
h)	GUI	q)	ALT
i)	POST	r)	DEL

4) What is the difference between RAM and ROM?
5) Explain the difference between cold boot and warm boot.
6) What is the basic function of DOS?
7) What does the booting process mean?
8) What is the maximum length a DOS 6.22 filename can have with its extension?
9) What is the difference between DOS 6.22, Windows 3.1, and Windows 95?
10) What are the characters that you cannot include as part of a filename?
11) What do the Caps and Num lock keys on the keyboard do?
12) What program initiates the booting process?
13) What is the difference between the internal and external DOS commands?
14) When are the CONFIG.SYS and AUTOEXEC.BAT files executed?
15) What is the difference between a .BAT file and an .EXE file?
16) Where are the internal DOS commands located? External DOS commands?
17) Which of the following filenames' syntax is wrong? Why?

a) TEST	b) TEST.HTML	c) SOFTWARES.DO	d) HARD\SOF.TXT
e) A.B	f) .BAT	g) EXPER12.EXT	h) TEXT HTM

18) What DOS commands can you use to create batch files?

19) How can you pause a batch file while it is running and make it display the comment "Press any key to continue"?

20) Within a batch file, how can you execute another batch file called TEST.BAT?

21) What is the importance of the CONFIG.SYS file? Is it required to boot up the system?

22) What is the importance of the AUTOEXEC.BAT file? Is it required to boot up the system?

23) What command do you need to install a device driver in a CONFIG.SYS file?

24) Explain the difference between the BUFFERS and FILES commands.

25) What do the commands DEVICE=C:\DOS\HIMEM.SYS and DOS=HIGH do in the CONFIG.SYS file?

26) What does the command BREAK=ON in the CONFIG.SYS file do?

27) Why is it not recommended to assign BUFFERS and FILES to their highest values and leave the BREAK command always ON in the CONFIG.SYS file?

28) What is the importance of the command PATH in the AUTOEXEC.BAT file?

29) How can you have Windows 3.1 immediately launched once the system is cold or warm booted?

30) What is the disadvantage of placing a lot of directories in the PATH statement?

31) What does the command PROMPT=PG in the AUTOEXEC.BAT file do?

32) How can you bypass the CONFIG.SYS and AUTOEXEC.BAT files when executing?

33) Explain the two major areas of computer memory.

34) What is the difference between dynamic and static memory?

35) What are the responsibilities of each of the following memory areas?

a) Upper memory	b) Conventional memory	c) Expanded memory
d) Extended memory	e) Virtual memory	

36) How big is the conventional memory?

37) How do you free up the conventional memory?

38) How can you find out how much space the conventional memory has left?

39) Why does having more extended memory in your computer speed up operation?

40) How does cache memory operate?

41) Why is it important to free up conventional memory?

42) What is the use of an EMM software?

43) Where is the ROM BIOS software loaded?

44) Why is it important to leave 20% of your hard drive empty?

45) In what memory does the operating system load when you cold or warm boot the system?

46) Why is it important to have the command HIMEM.SYS be executed in the first line of the CONFIG.SYS file?

47) What is the difference between extended and expanded memory?

48) Where does Windows 3.1 place the temporary files it creates when it is first launched?

49) What are real mode and protected mode?

50) What does NT stand for in Windows NT?

51) List the system requirements for Windows 95/98/NT and 2000?

2

Configuring and Upgrading PC Hardware

Regardless of what type of PC you have your computer system houses the same components. Your computer system houses the PC's disk drives, central processing unit, memory, and other hardware devices, such as a modem or a sound card. When you perform hardware upgrades, you will need to open your PC system unit. To open your system unit, you should turn off the power and unplug the system.

Opening up the PC

To remove the system unit cover, you have to remove screws that are normally found on the outer edge of the system. Gently slide the PC cover off. If the cover does not come off easily, do not force it. The system unit contains many cables that may be tangled. Whether you have a tower or a desktop PC, the components within a system unit are the same. Before you do anything, you must touch the outside of your system unit chassis to ground yourself. This will greatly reduce the possibility of damaging your computer's chips due to static electricity. You might want to purchase a clip that grounds your computer while you work.

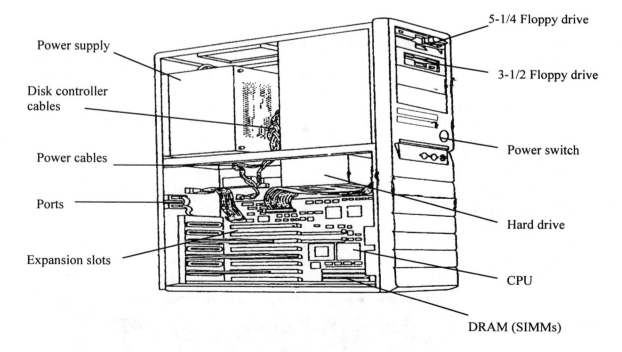

Figure 2.1 Inside the PC

The computer's motherboard is a large electronic board that contains the majority of your computer's chips. The motherboard contains the CPU, BIOS, and PC memory (SIMMs or DIMMs). The motherboard is the most expensive component of your computer. You must treat it with care when you need to replace it. Inside the PC you will also find a power supply. The power supply is there to change the alternating current from your outlet at home into direct current that your computer can use. The power supply also contains a fan that helps cool the PC. Also in the system, the hard and floppy drives are connected to the power supply and have cables that connect them to the computer. Expansion cards, such as a fax modem, fit into expansion slots on the motherboard to enhance the computer's capabilities.

Computer Ports

A port lets you connect a device to your PC. Ports let you connect devices such as a monitor, a mouse, a printer, or even a keyboard. If you examine your PC's ports, you will note that some ports are designed to plug into a cable, while others are designed for a cable to be plugged into the port. Ports and cables, therefore, are classified as male or female, based on whether they plug in or are plugged into. A male cable or connector contains visible pins that plug in. A female cable or connector has holes rather than pins. You can change a port's gender using a gender changer. In other words, you can change a male into a female, or vice versa.

Figure 2.2 Various computer ports

Parallel Ports

You can transmit 8 bits of data at a time over 8 wires through a parallel port. The maximum length of a parallel cable is about 10 feet. It has a female port with 25 holes that connect to a 25-pin male connector. The 25 holes on the port are as follows:

Hole 1 = strobe signal
Holes 2 to 9 = data bits
Hole 10 = acknowledge
Hole 11 = busy
Hole 12 = paper out
Hole 13 = select
Hole 14 = auto feed
Hole 15 = error
Hole 16 = initialize printer
Hole 17 = select input
Holes 18 to 25 = ground

Hole 1   Hole 25

Figure 2.3 25-hole parallel port

Serial Ports (RS232 Ports)

Serial ports can connect devices such as printers, modems, a mouse, and more. Serial ports are slower than parallel ports, but can have a cable with a maximum distance of 50 feet. Serial ports transmit data one bit at a time on a single wire. A PC can support up to four serial ports, named COM1, COM2, COM3 and COM4. However, you can only use two serial ports at a time. This is because COM1 and COM3 use the same IRQ (Interrupts request) number, which is IRQ4. COM2 and COM4 use IRQ3. As you will see in the next section, normally no two devices can share the same IRQ. But here we have a special case: two COM ports may have the same IRQ number as long as they do not communicate at the same time. If they do the computer will lock up. Serial ports come in either 9- or 25-pin ports. Remember, a COM and an LPT port is nothing more than a preset combination of an IRQ and I/O address.

Figure 2.4 25-pin and 9-pin serial ports

Keyboards

Keyboards are the most common way of getting information into a PC. A computer keyboard consists of keys that, when tapped, send a message to the CPU telling it what key is depressed, or when the key is no longer depressed.

Keyboard Ports

Keyboard ports are usually circular. Make sure to align the cable's pins to the port before you plug in the cable. All types of keyboards operate on capacitive keyboard technology. All the pads of the keyboard are scanned for current changes every few microseconds. The instant current flows, a keystroke can be detected. The microprocessor is always scanning the keyboard to see if a key has been pressed. Once a keystroke has been detected, the microprocessor built into the keyboard generates a scan code that indicates which key was pressed. The scan code is then converted to serial data and relayed to the CPU in the computer. Each press of a key generates two different scan codes—one when the key is pushed and one when the key is released. Keyboard scan codes are transmitted serially on a single wire. The keyboard cable has a 5-pin connector to fit into a 5-hole port.

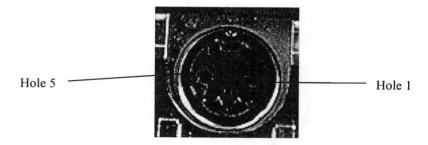

Figure 2.5 5-hole keyboard port

The pins on the connector are as follows:

Pin 1 = keyboard clock
Pin 2 = keyboard data
Pin 3 = keyboard reset signal (normally not used)
Pin 4 = ground
Pin 5 = + 5 volts DC

Mouse

A mouse is a pointing device used for data input. It is simply a device that moves a cursor to any area of the monitor screen. Two kinds of mouse interfaces are available: a serial mouse which plugs into the serial port of the PC, and a bus mouse which comes with an adapter card and plugs into the expansion slot inside the PC. Depending on your monitor type, the monitor port will connect to either a 9- or 15-pin port and your printer port can connect a printer to a serial or a parallel port named LPT1, LPT2, or LPT3. All PCs can support up to three parallel printers.

To connect devices to your computer such as a printer, you must connect a cable from the device to a port on the back of the PC. You must be able to differentiate between male and female connectors and know how to use a gender changer. You must also understand the purpose of common ports and how to add ports to your computer. When you add new hardware to your computer, you must be able to determine if your PC needs additional ports or not. Devices such as a mouse, a printer, and a keyboard all connect to ports on the back of your PC.

Looking at the back of the PC, you will notice several different types of ports. Some are designed to plug into a cable and others are designed for a cable to be plugged into the port. Ports and cables are, therefore, classified as male or female, based on whether they plug in or are plugged into. A male connector or cable contains visible pins that plug in, whereas a female connector or cable has pinholes. To help you connect a cable to a port, ports and cables are shaped in such a way to ensure that you plug in the cable correctly. Common ports come in 25-pin and 9-pin varieties. You should purchase male cables for female ports, and vice versa. If you purchase the wrong cable type, you can use a gender changer to change the port's gender.

The keyboard port is a 5-pin circular port. Like all cables, you must align the keyboard cable to the port openings before you can connect the cable. The VGA monitor connects to a 15-pin port. You can connect a printer to either a parallel or a serial port. Parallel ports transmit eight bits of data at a time, while serial ports transmit one bit of data at a time. Most PCs come with 2 serial ports. The advantage of a serial port is that it can transmit data over a longer distance than a parallel port. Parallel ports usually use 25-pin female connectors, while serial ports use either a 25- or 9-pin connector. Most serial ports use a male connector. A serial mouse port connects to a 9-pin serial female connector, while a bus (PS/2) mouse connects to a port that looks like a small version of your keyboard cable.

Figure 2.6 PS/2 mouse port

CPUs

The Central Processor Unit (CPU) or the microprocessor on the motherboard is the brain of your computer. The CPU is a chip that actually processes instructions and carries out commands. Every other part of your computer serves the microprocessor. There are several CPU attributes that one must understand:

Clock Speed

Inside every computer is a quartz crystal, called the internal clock, that sends out an electronic pulse, with each pulse called a cycle. The clock's beat extends to and controls the processing rate of nearly every chip in the computer. In the original XT, the clock emitted an electronic signal with a frequency of 4.77 million cycles per second. Internal clocks in computers today emit a signal with a frequency of 200–500 million cycles per second. The term hertz is another way of expressing cycles per second. Megahertz (MHz) means millions of cycles per second.

The more quickly the clock beats, the more quickly the CPU processes information. The CPU breaks down each instruction in a program into a series of steps. One step is completed each clock cycle. So, the faster the clock ticks, the faster the instruction is completed. You can use the megahertz rating of a CPU chip to compare performance only within a single family of chips. A system with a 200 MHz Pentium chip, for example is faster than a system with a 100 MHz Pentium chip, because the 200's clock ticks 100% faster.

All of the chips on the system board are linked together by a series of printed circuit lines known as the bus. A bus is a shared path along which all the parts of the computer are connected. Every control chip, byte of memory, and expansion board is connected directly or indirectly to the bus. Data is passed between components along the bus. We can access the bus through the expansion slots on the system board. The bus is divided into four sections: power, address, data and control.

Data Bus

The data bus is a group of wires used to carry data through the system. The data bus is broken into two sections: internal and external. The internal data bus (also called register size) is a special memory location within the CPU that holds the numbers being worked on. The larger the registers are, the larger the numbers that can be handled. If we have a 16-bit register, then the largest number that can be handled by a single ADD instruction is 32,768. In a CPU with a 32-bit register, that number grows to 2,147,463,684. So the 32-bit CPU chip needs only one instruction to add a large number, while a 16-bit CPU takes 4 or 5 instructions and correspondingly more time.

Like a highway carrying cars between cities, the external data bus carries information to and from the various components on the system board. Like a highway, the wider it is, the faster information will flow. The 8088 CPU chip had an external 8-bit data width, similar to a one-lane highway, which meant that every number going in or out of a 16-bit CPU register had to be processed in two steps. The 8086 and 80286 CPUs have a 16-bit external bus, (equivalent to a two-lane highway), and so data flows more quickly. The 80386 and 80486 have a 32-bit external data bus, which is equal to a four-lane highway. The Pentium, Pentium Pro, Pentium II, and Pentium III have a 64-bit external data bus or an eight-lane highway.

Address Bus

The address bus works along with the data bus. It is a group of wires used to transmit the addresses of the memory locations or devices that are to receive the data that follows. The maximum amount of memory (in bytes) you can have your computer is 2 raised to the power of (number of lines). So an 8088 with its 20 address lines can have a maximum of $2^{20} = 1,048,576$ or 1 Megabyte (MB). The 80286 with 24 address lines can have $2^{24} = 16$ MB. The control bus works for the address bus and is a group of wires that tells the CPU when its receiving or sending data. The power bus is a group of wires for supplying electrical power to the CPU.

Cache Memory

Although the speed of CPUs has increased dramatically in recent years, the speed of memory has not. RAM memory has become one of the bottlenecks (points of congestion) in today's high-performance system boards. CPU chip manufacturers now include some very high-speed memory, known as cache memory, on the CPU chip. The larger the cache, the better the CPU's overall performance will be. It consists of static memory chips used to hold operating instructions and data likely to be needed next by the CPU.

There are many types of Intel microprocessors used by IBM PCs and compatibles as seen in Table 2.1.

80486

Introduced in 1989, the 80486 microprocessor was Intel's new and improved 80386. In fact, from a software standpoint, the 486 is almost indistinguishable from the 386 except the speed. All the microprocessors before the 80486DX had an optional companion chip to do arithmetic calculations. This optional companion chip is called a math coprocessor. The addition of a math coprocessor on board will speed mathematical calculations up to 100 times faster with a high degree of accuracy. The 80486DX2 doubled the speed of the internal CPU speed, while the other chips on the motherboard use a slower clocking rate. The DX2 chip gave board designers a low-cost way to create faster system boards without major changes.

Figure 2.7 80486DX and 80486DX2 CPUs

CPU Type	Clock Speed (MHz)	Internal Data Bus	External data bus	Address bus	Max RAM Access	Internal Cache level 1 / Level 2	No. of Transistors on board
8088 (XT)	5–10	16-bit	8-bit	20-bit	1 MB	No / No	40,000
8086	5–10	16-bit	16-bit	20-bit	1 MB	No / No	40,000
80286	6–20	16-bit	16-bit	24-bit	16 MB	No / No	140,000
80386SX	16–40	32-bit	16-bit	24-bit	16 MB	No / No	275,000
80386DX	16–40	32-bit	32-bit	32-bit	4 GB	No / No	275,000
80486SX	16–43	32-bit	32-bit	32-bit	4 GB	8 KB / No	1,185,000
80486DX	20–50	32-bit	32-bit	32-bit	4 GB	8 KB / No	1,200,000
80486DX2	50–66	32-bit	32-bit	32-bit	4 GB	8 KB / No	1,200,000
80486DX4	75–100	32-bit	32-bit	32-bit	4 GB	8 KB / No	1,200,000
Pentium (586)	60–200	32-bit	64-bit	32-bit	4 GB	16 KB / No	3,000,000
Pentium Pro (686)	133–266	32-bit	64-bit	32-bit	4 GB	16 KB / 256 KB	CPU = 5,500,000 Cache = 15,500,000
Pentium II (786)	233–450	32-bit	64-bit	32-bit	4 GB	16 KB / 512 KB	CPU = 5,500,000 Cache > 30,000,000
Pentium III (886)	450–1000	32-bit	64-bit	32-bit	4 GB	16 KB / 512 KB	CPU > 5,500,000 Cache > 30,000,000

Table 2.1 Intel's CPU types

Pentium

Intel named the 80586 CPU the Pentium for two reasons—the numbers, 286, 386, and 486 were ruled by courts to be in the public domain, and the internal design of the chip was designed with five sides. Intel chose another name to protect its trademark rights. The Pentium is fully instruction compatible with earlier 8088, 286, 386, and 486 processors. The Pentium CPU chip contains more than 3 million transistors on a single chip, and is two to three times more powerful than the 486. It also has more internal cache, 16 KB cache instead of 8. The Pentium has 64-bit data bus width, but still has a 32-bit internal register size like the 386 and 486 processors.

MMX (Multi Media Extensions)

The increases in Internet access with its highly graphical web pages, and the increased complexity of computer games, have changed the nature of our requirements from a CPU chip. Intel's MMX is a series of CPU enhancements designed to speed up multimedia operations. Through the addition of 57 new instructions, the CPU can now work on audio and video data 64 bits at a time, as opposed to previous CPUs, which could operate on only eight bits at a time. The MMX CPU also increases on-board cache memory (L1 cache) to 32 KB, which boosts the performance of all instructions.

Pentium Pro

All the previous processors were designed to be backward compatible while improving the performance of older 16-bit operating systems and programs. The Pentium Pro will be able to run 16-bit programs, but it is designed to optimize the performance of 32-bit programs, operating systems, and network operating systems. The Pentium Pro is a Multi-Chip Module (MCM). It is a dual-cavity design with the larger CPU on the right side and the 256K L1 (level 1) cache on the left. The transistor count for the CPU is 5.5 million, while the cache is 15.5 million.

Figure 2.8 Pentium Pro (MCM), CPU on the right and L2 cache on the left

This unique packaging form allows engineers to construct high-performance system boards that will be speedy and inexpensive. The L2 (level 2, external) cache memory is a set of static RAM chips on the motherboard, forcing the CPU to communicate with the cache memory through the external bus at a reduced rate of 66 MHz. By combining the two chips together with their own high-speed bus, access to the L2 memory is at full processor speed, not the reduced external bus speed.

Intel's Pentium Pro uses the term Dynamic Execution to describe a combination of multiple branch prediction, data flow analysis, and speculative execution. These techniques are used to pre-fetch the contents of memory locations before they are actually requested. While the CPU is busy executing an instruction, normally idle circuitry can access memory, saving time when the data is actually needed. The Pentium Pro has a 14-stage pipeline, compared to the Pentium's 5 stages. This permits more instructions to be overlapped for better performance. Each stage has been optimized. The Pentium Pro has many other advantages, such as three-way pipelines which permit the simultaneous execution of up to five instructions, a closely coupled secondary cache, a transactional I/O bus, more execution units, and more.

Pentium II

Pentium II and III CPUs are faster than a Pentium Pro with MMX. They are equipped with a new SEC cartridge replacing the old-style SPGA socket of the Pentium Pro and allowing more room on the motherboard. Both chips a have high capacity of L2 cache for better performance. Although the Pentium II is a great CPU, it has some limitations. For example, they generate the greater-than 200 MHz speeds by using large multipliers of 66 MHz bus clock, while Cyrix and AMD CPUs (the chief rivals to the Intel CPUs) currently handle bus speeds approaching 100 MHz.

Figure 2.9 Pentium II

Pentium III

The Pentium® III processor at 1 Gig, 866, 850, 800, 750, 733, 700, 667, 650, 600, 550, 553, 500, and 450 MHz extends processing power further by offering performance headroom for business media, communication, and Internet capabilities with 100- and 133-MHz bus systems. The Pentium III processor offers great performance for today's and tomorrow's applications as well as quality, reliability, and compatibility.

The Pentium III processor integrates the best features of the P6 micro-architecture processors—Dynamic Execution performance, a multitransaction system bus, and Intel MMX media enhancement technology. In addition, the Pentium III processor offers Streaming SIMD (Single Instruction Multiple Data) Extensions—70 new instructions enabling advanced imaging, 3D, streaming audio and video, and speech recognition applications. To better experience the power of the Pentium III processor, combine it with technologies from Intel's Labs including Intel Streaming Web Video software and the Intel videophone.

Cyrix / IBM 6x86

This CPU is a Pentium clone that offers high performance at a low price. Through the use of innovative, sixth-generation architectural techniques, the 6x86 processors can match the performance of faster CPUs from other manufacturers. Cyrix has found that their processor performs as well as faster Pentiums, so they have adopted a slightly different naming convention based on performance. The 6x86-PR200+ actually runs at 150 MHz, the PR166+ runs at 133 MHz, and so on. The 6x86 CPU is also produced under the IBM model.

The Cyrix 6x86 processor achieves high performance through the use of two optimized super-pipelined integer units and an on-chip FPU (Floating Point Unit). The integer and floating point units are optimized for maximum instruction throughput by using advanced techniques including register renaming, out-of-order completion, data dependency removal, branch prediction, and speculative execution.

Figure 2.10 Cyrix 6x86MX PR233

AMD 5x86 / K5 /K6

AMD (Advanced Micro Devices) produced 5x86 versions in speeds up to 133 MHz. AMD's Am5x86-P75 offers performance similar to Intel's Pentium-75 and Cyrix Cx5x86-100. AMD-compatible CPUs are called the K5 series. They are designed to be compatible with "socket 7" (Pentium) type sockets on Pentium motherboards. This socket compatibility enables manufacturers to minimize redesign costs by requiring minor BIOS modifications for existing motherboards. CPU speeds of 75—166 MHz are available.

AMD's sixth-generation microprocessor is known as the AMD K6 processor, and is designed to compete with Intel's Pentium Pro. This processor contains a fully compatible implementation of the MMX instruction set. The AMD-K6 processor retains the "socket 7" system bus and electrical compatibility.

The following are steps that should be followed when inserting a CPU into a socket:

- Don't touch the pins: you might destroy the CPU.
- Notice the "orientation notch" or "index notch" at the lower-left-hand corner of the CPU. It is designed to help you align the CPU correctly. It must align with the notch on the socket.
- Be careful: improperly installing your CPU will almost certainly destroy the CPU and/or the motherboard.

Motherboards

The most important part of your system is the motherboard. The motherboard, sometimes referred to as the system board, is the foundation of your computer. It supplies all of the electrical connections between components such as the CPU, ROM BIOS, expansion slots, and support circuitry for the power supply and RAM.

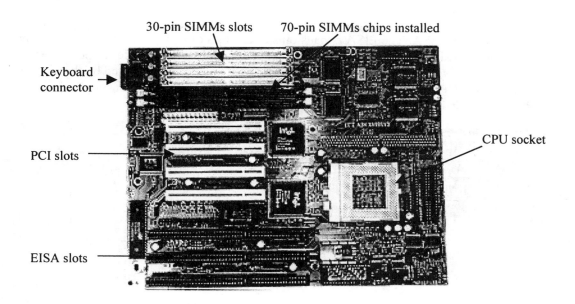

Figure 2.11 Typical motherboard layout

Expansion Slots

Expansion slots on the motherboard allow you to plug in electronic boards to enhance the operation of your system. The slots allow you to add boards such as a serial or a parallel port, a video card, a mouse, or a fax modem. Understanding how types of expansion slots differ with various types of motherboard architecture will help you understand the difference between various computers.

ISA

This stands for Industry Standard Architecture. The ISA expansion slot is an 8-bit slot with a 62-pin connector as shown in Figure 2.12. It provides eight data lines and twenty address lines, which enables the board to reside with the 1-MB conventional memory.

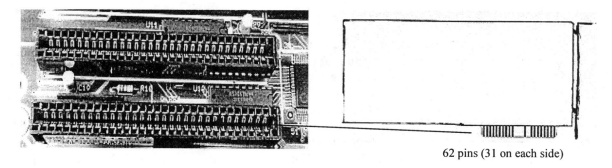

62 pins (31 on each side)

Figure 2.12 8-bit ISA slot and card

EISA

This stands for Extended ISA. The EISA expansion slot consists of an ISA slot with an additional 18-pin connector extension slot to support a 16-bit board. 32-bit ESIAs were developed to meet the demand for greater speed and performance from expansion peripherals incited by the speed of the 80386 and 80486 CPUs. Unlike the MCA bus, the 16-bit and 32-bit EISA ensure backward compatibility with existing ISA peripherals.

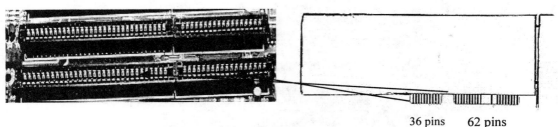

36 pins   62 pins

Figure 2.13 16-bit EISA slot and card

MCA

This stands for Micro-Channel Architecture. The ISA bus was considered "open architecture;" that is, anyone can use it without paying royalties. MCA, on the other hand, is a bus standard for IBM PS/2 systems, and anybody who wants to use it must pay a royalty to IBM. Remember that expansion boards that fit in an ISA slot will not fit in an MCA slot. One major flaw with an MCA bus was its lack of compatibility with ISA; none of the old cards would work in an MCA system.

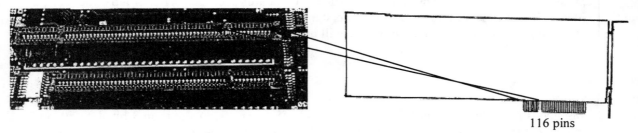

116 pins

Figure 2.14 32-bit MCA slot and card

Local Bus

The local bus was designed to speed up video transfers. The earliest designs actually consisted of video circuitry on the motherboard as opposed to board-mounted slots. The ISA bus presents the worst barrier to a CPU's ability to get data to the video card; upgrading to an EISA or MCA bus could improve the flow. The only problem with EISA or MCA buses is the high cost of the systems and video cards needed to take advantage of them.

The ultimate solution is to bypass the bus altogether and tap the CPU chip's local bus; this approach allows extremely high-speed transfer of information between your CPU chip and the display adapter. To implement this technique, however, you'll need to have the display adapter's circuitry built directly onto the computer's motherboard or into a special direct-access slot called a local bus slot. If you are a Windows demon, you should definitely consider local bus video. Local-bus technology breaks the bottleneck by bringing peripherals like video controllers up to the clock speed. When it comes to computer graphics, it is faster to use the local bus than the expansion bus. On a 33-MHz 486 local bus PC, a video controller could match the 33-MHz clock speed of the processor.

VESA Local Bus

VESA (Video Electronics Standards Association) is an industry association that was formed in 1989 by hardware and software developers for the purpose of standardizing the interaction between software drivers, display boards, monitors, and applications. The VESA Local Bus standard was formally announced on August 28, 1992 to solve the compatibility problems several proprietary local bus designs had.

The VESA local bus standard is only available for 486 systems, and has been supplanted by the PCI bus architecture for all Pentium, Pentium Pro, Pentium II, and Pentium III systems. There is little reason to purchase a new system with VLB slots. The VL-Bus connector is a standard 116-pin 32-bit micro-channel-type connector and is located in-line with a system I/O bus connector and can be implemented on ISA, EISA, and MCA system boards. This layout allows full use of all system I/O bus slots if VL-Bus add-in boards have access to all system I/O bus features.

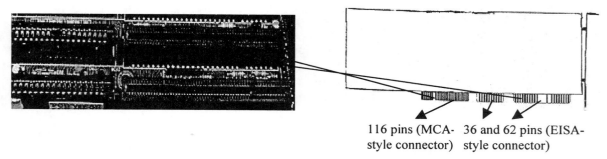

116 pins (MCA-style connector)   36 and 62 pins (EISA-style connector)

Figure 2.15 32-bit VESA local-bus slot and card

PCI

This stands for Peripheral Component Interface. It is a type of a local bus, but PCI is the decoupling of the processor and main memory subsystems from the standard expansion buses (ISA, EISA, and MCA). PCI specification is designed to accommodate faster CPUs without requiring constant peripheral upgrades and redesigns. PCI is best suited for Pentiums and any other 64-bit CPU chip. Unlike the VL-Bus, which has both an ISA and VL-Bus connector in line, a PCI bus slot has a single connector (which looks like an MCA slot, but isn't MCA-compatible).

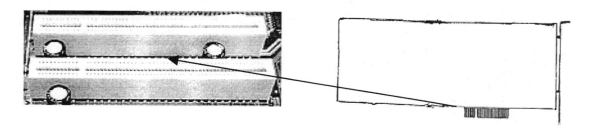

Figure 2.16 32 / 64-bit PCI slot and card

PCMCIA

This stands for the Personal Computer Memory Card International Association. It was established in 1990 to set standards for credit-card-size memory modules. This was due to the explosion in the use of portable computing.

To install an expansion board, you must be sure to read the manual and any instructions accompanying the board. To prevent any static damage, leave the board in its protective bag until you take it out to set the jumpers as shown in the following steps:

- Turn off the computer's power, and remove the case cover.
- Use a wrist strap, or at least touch the power supply casing.
- Remove the board from its bag, and set any jumpers or DIP switches.
- Insert the board into an empty slot, tighten the retaining screw, and attach the cables (if any).
- Run any diagnostic program that comes with the board.
- Install any driver software.
- Record the jumper or DIP switch settings; this will make life easier the next time you work on this computer.

Test each and every device in the system, looking for any conflicts. If everything checks out, close up the computer case. There are also some ROM (Read Only Memory) chips on the motherboard that contains commands your computer needs to get itself going. The ROM is non-volatile, meaning that the content of ROM is retained even when the power is off. The instructions in ROM allow the computer to get started when the power is switched on. You should follow the following steps when replacing a motherboard:

- First, you must remove the old motherboard by removing all the cards.
- In addition, remove anything else that might impede removal or installation of the motherboard, such as a hard or floppy drive.
- Keep track of your screws. The best way to do this is to temporarily return the screws to their mounting holes until the part is to be reinstalled.
- Document the position of all wires, especially the little ones such as speaker, turbo switch, turbo light, and reset button in case you need to reinstall them.

BIOS Memory Chips and Replacing Its Battery

ROM BIOS

Because computers differ in the number or type of disks, video displays etc., the battery-powered CMOS (Complementary Metal Oxide Semiconductor) memory chip, sometimes called the BIOS (Basic Input Output System), on the motherboard is used to help the computer keep track of its configurations. The BIOS chips are a collection of software routines between the hardware and the systems software. These routines control the input and output of data. They include device drivers, utilities, interrupt service routines, and other code to move data between the system hardware and the systems software. The term ROM BIOS refers to those routines that are permanently stored into a ROM chip. The ROM BIOS is made up of the following:

- POST (Power-On-Self-Test)—This is a built in self-diagnostics program that runs each time the system cold boots.

- BIOS setup utility—This allows you to store system configuration values such as the number and type of disk drives, amount of memory, and the presence of optional components.
- BIOS routines—These are the input and output subroutines that interface between the hardware and software.
- Bootstrap program—This program initializes the chips and adapter cards. It sets up the interrupt vector table in low RAM. It finds out what optional equipment is in the system. Finally, it boots the operating system.

Figure 2.17 Typical BIOS chip

The computer will never be aware of any hardware upgrades until you update the BIOS memory. As you know already, the information in RAM is always lost when the power is turned off. For the PC to remember the key settings, the PC uses the battery-powered BIOS memory. You can access your system's BIOS settings during the self-test when you first turn on the PC. To access the BIOS, you must press one of the following keys or key combination when your system completes its memory count during self-test: <Del>, Ctrl>-<Alt>-<Enter>, <Esc>, <Ctrl>-<Alt>-<Ins>, or <Ctrl>-<Alt>-<Esc>. Depending on your system, it might display a menu of options. The menu options will have a range of settings like current date, time, disk types, etc. Do not forget that it is very important to specify the cylinders, heads, sectors, and size of your hard drive in BIOS.

A 12-volt battery powers the BIOS. This battery will eventually fail. You must restore all the settings back into BIOS after replacing the battery. That is why you should write down all your current system settings and store them in a safe place. As long as the system starts successfully, you can ignore the BIOS settings, but if you receive the message "Invalid System Settings-Run Setup," the BIOS battery probably has failed and you must change the battery.

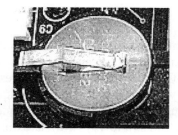 

Figure 2.18 Typical DC batteries

You can change the battery following the steps listed below.

- Turn off and unplug the computer.
- Open the system unit.
- Locate and replace the battery.

- Close the system unit.
- Plug the power back in and cold boot.
- Restore the previous BIOS settings.

The BIOS are a pair of chips on the motherboard that are responsible for all the input and output operations. If you have an old PC, you might have to upgrade the BIOS chips before you can use a newly installed device. It is a good idea to upgrade the BIOS on an old system to improve system performance, but before you upgrade the BIOS chips, you must check with the PC manufacturer first to ensure compatibility. You should never upgrade the BIOS to gain better system performance; instead, you should upgrade the BIOS to gain support for new devices. Make sure to place the odd and even chips in the correct sockets, or your system will not start. Here are the steps you should follow to upgrade the BIOS chips:

- Turn the power off and unplug the PC.
- Make sure that you are static free.
- Remove the unit cover.
- Locate the BIOS chips on the motherboard.
- Remove the chips using a chip extractor.
- Insert the new BIOS chips into the motherboard.
- Replace the unit cover, plug the power in, and turn your system on.

Flash BIOS

In anticipation of changes needed to support new features to PCs such as Plug-and-Play (PnP), bootable CD-ROM drives, and high-capacity floppy drives such as ZIP drives, many computer manufacturers have been using "Flash BIOS", or as it is technically called, "EEPROM" (Electronically Erasable and Programmable Read Only Memory). Under program control, a high voltage is sent to the chip, causing it to erase its contents, followed by the new information to be loaded into the ROM. The information will remain in the chip until such time as it is erased. Flash ROM updates are received as program files, either on diskette from the manufacturer, or more commonly, downloaded from a web site on the Internet.

IRQs and DMA

IRQs

When a device such as a mouse or a modem requires attention, it sends an interrupt request (IRQ) to the CPU. The CPU will then stop executing the task it is performing and devote its time to performing special instructions for that device. Once the CPU completes the task, it resumes executing the previous task. Each piece of hardware that you install will require a unique IRQ number, a memory buffer address, and sometimes a Direct Memory Access (DMA) address. A 286 CPU or higher can accommodate 16 different IRQ settings (0 to 15). Two hardware components cannot have the same IRQ number. If they do, the system will not operate correctly. Remember that if you set a device to a COM/LPT port you are using an IRQ. This is always a big problem for new technicians who do not understand IRQs and their relationship to COM/LPT ports. If someone has a COM1 port and then tries to install some other devices to IRQ4, the system will lock up. Using the Device Manager in Windows, you can view all 16 IRQs and their respective addresses. Below is an example of an IRQ status using the Device Manager in the Control Panel.

IRQ	Address	Description	Detected	Handled By
0	1A68:1875	Timer Clock	Yes	Block Device
1	1A68:1923	Keyboard	Yes	Block Device
2	F000:F853	Second 8259A	Yes	BIOS
3	F000:F853	COM2: COM4:	COM2:	BIOS
4	210E:3C56	COM1: COM3:	COM1: Serial Mouse	MOUSE.COM
5	F000:F853	LPT2:	Yes	BIOS
6	0401:00B7	Floppy Disk	Yes	Default Handlers
7	0070:06F4	LPT1	Yes	System Area
8	0401:0052	Real-Timer Clock	Yes	Default Handlers
9	F000:F847	Redirected IRQ2	Yes	BIOS
10	06DB:0218	(Reserved)		MOUSE.COM
11	F000:F853	(Reserved)		BIOS
12	F000:F853	(Reserved)		BIOS
13	F000:F84A	Math Coprocessor	Yes	BIOS
14	0401:0117	Fixed Disk	Yes	Default Handlers
15	F000:F853	(Reserved)		BIOS

There are 2 interrupt controllers on the motherboard to handle interrupt requests for the CPU. The first one handles IRQs 0 to 7 and the second handles IRQs 8 to 15. The PC takes IRQ 2 space to access the second set of interrupts. In the above list the 8259A is an interrupt request chip. IRQ 2 is therefore not really used. Any hardware that has an IRQ 2 setting is routed to IRQ 9, where you can use the second set of IRQs. In addition, any IRQ that is available will have a "Reserved" entry in its description. Hardware components usually come with software, jumpers, or DIP switches to set an IRQ setting. Make sure that the IRQ on the card is not the same as the IRQ of another hardware on your system. In addition to IRQ numbers, some cards may require unique I/O and memory addresses reserved for their use. Again those I/O and memory addresses must not be the same as another component's address.

When installing a new expansion card, you must make sure that its settings are unique. To ensure that they do not conflict with another card already in your computer, use one of the following approaches:

- Open up the computer and look at the settings on each card in the slots.
- Keep notes of settings as you install cards on a sheet of paper kept in a filing cabinet, or kept in the computer itself.
- Run one of the configuration programs that display the IRQs in use.

DMA

DMA (Direct Memory Access) is a way for the expansion board to transfer its data to/from the system board's RAM without CPU intervention. Only after the data has been transferred will the interrupt occur: at that time the data is ready to be worked on. Direct memory access allows a device to stuff data into your computer's RAM, bypassing the microprocessor. When an external device wants to feed data into your PC's RAM, it triggers its assigned channel. Only when the microprocessor sends back an acknowledgment can data transfer begin. In this way, your PC isn't bombarded with data coming from all directions at once.

A channel is a connector on the expansion slot on the PC's main motherboard. It is where data always takes the same route between the external device and RAM chips.

DIP Switches and Jumpers

DIP Switches

DIP switches are banks of one or more miniature on/off switches. They are called "DIP" because of their design: "dual in-line package," the same design as an ordinary integrated circuit. The switch levers are very small, and generally must be set using a sharp instrument, such as the tip of a ball-point pen. Push the level to "On" (or "1") to enable the switch. There's generally no marking for the "off" setting. (Avoid using a pencil to set DIP switches; the graphite from the pencil can come off and impair the electrical contact inside the switch.)

The more hardware boards you add to your computer, the more you will probably need to change a jumper or a DIP switch setting. You need to change the settings to avoid IRQ conflicts with existing boards on the PC. As you can see from Figure 2.19, a DIP switch acts like light switch. DIP switches are usually labeled On/Off or 1/0.

Figure 2.19 Typical DIP switches

Jumpers

Jumpers serve the same purpose as DIP switches, but because they are cheaper for the manufacturer to implement, they are far more common. The jumper consists of two pins mounted on the circuit board of the hardware. You place a shunt or jumper block over the two pins to establish the electrical connection. If you remove the jumper block, the connection is broken. You can use a small pair of needle-nose pliers or blunt tweezers to install and remove jumpers. Be very careful not to bend the connecting pins. They can be broken off if mistreated. *You should never change a jumper or a DIP switch setting without writing down the original setting.* This way you can always restore the original setting whenever you need to.

Jumper placed   Jumper removed

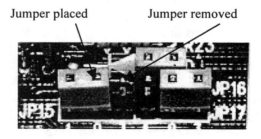

Figure 2.20 Typical jumpers

Memory

Adding more memory to your PC will improve system performance significantly. Every program that needs execution must be transferred from the hard disk to RAM memory. As programs grow bigger, they require more RAM. Furthermore, if you use Windows or have several different programs running at the same time (multitasking), they will each require their own memory space. When the PC memory runs out of space, it has to swap programs in and out of the hard drive to let several programs run at the same time. The disk swapping slows down performance significantly. One of the best ways to improve system performance is to add more memory, which in turn reduces swapping between the hard drive and memory. You will probably see a major improvement in performance after you add memory if you are running multiple programs at the same time.

Before you add any memory to your PC, you must know how much memory your system has. Executing the command MEM will display how much memory your system has. The extended memory (RAM) is located on the motherboard and it uses an upright chip called a SIMM.

SIMM

This stands for Single Inline Memory Module. SIMMs consist of several memory chips stored on one module. SIMM chips come in the sizes 1 MB, 2 MB, 4 MB, 8 MB 16 MB, 32 MB, 64 MB, and so on. Most PCs can add memory up to 128-MB chips. In most cases, the memory you add to the motherboard must match the specification of the memory currently installed. All memory chips are classified by their size and their access speed. The access speed is expressed in nanoseconds (ns), which are billionths of second. The lower the number, the faster the chip will be. But remember that if you have a 4 MB 70 ns SIMM and add another 4 MB 90 ns chip, your system will run at the slower speed. In this case all 8 MB will run at the speed of 90 ns.

There are two types of SIMMs—30-pin and 72-pin. The 30-pin SIMMs have a 30 pins or contacts along the edge. The most important thing to know about the 30-pin SIMMs is that they have an 8 (1-byte) data-bit width. The 72-pin SIMMs have a 70 pins or contacts along the edge. They have a 32-bit data width. 72-pin SIMMs take up less space on the motherboard.

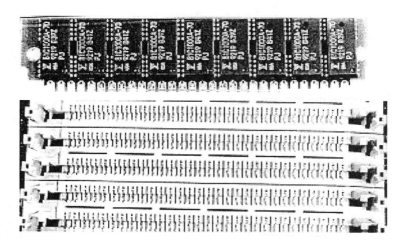

Figure 2.21 Typical 30-pin SIMMs and slots

You need to be very careful when you insert a SIMM chip into its socket. The SIMM chips are held in place by small notches at the end of the socket. You must insert the SIMM chip at an angle and then stand it upright. Make sure that the chip is inserted into the notches at the end of the socket. In some cases, you must inform your computer's BIOS that you have installed new memory, after you have finished the physical installation.

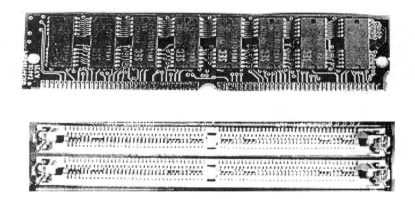

Figure 2.22 Typical 72-pin SIMMs and slots

DIMM

This stands for Dual Inline Memory Module. It has chips and contacts on both sides and is expected to be popular in the future, as we try to squeeze more memory into our computers.

DRAM

This stands for Dynamic RAM. It is the most common and cheapest type of memory chip. It is made up of millions of tiny capacitors. A capacitor is an electrical component that can hold an electrical charge used to represent a binary 0 or 1. An unavoidable property of the capacitor is the inability to retain that charge indefinitely. A DRAM capacitor can only hold the charge for a few milliseconds, and once discharged its value is forgotten (that is why everything in memory is forgotten when the power is turned off.)

The DRAM's capacitor cells must be refreshed continuously. That means that the data in each location must be read, and then rewritten to the same location, thereby recharging the capacitors. The motherboard contains circuitry to continuously refresh all DRAM locations so none of the contents are lost. The refresh cycle imposes a performance penalty on the computer, but we must make price/performance compromises to build affordable PCs.

EDO RAM

This stands for Extended Data Output and is known as hyper-page mode. It speeds up memory access by up to 30% by allowing the chip to load data while at the same time it switches to the next address, rather than waiting between these operations as DRAM does. Implementing EDO relatively expensive modification to ordinary DRAM control circuitry, so we can look forward to seeing it in many systems. The cost of EDO RAM is expected to drop to that of DRAM once production reaches a high level.

SRAM

This stands for Static RAM. Unlike DRAM, SRAM is made up of transistors and therefore does not require continuous memory refresh. SRAM is about four times fast as DRAM, which would make us want to construct a system with only SRAM. But SRAM chips are substantially larger and more expensive than DRAM making such a system impractical.

VRAM

This stands for Video RAM. Memory on video card has two masters, the CPU that wants to load information to be displayed, and the video card's RAMDAC (RAM Digital/Analog converter), which converts the digital bits into an analog RGB signal used by the monitor. When standard DRAM is used on the video card there is only one access port available, and both devices must share the one port. With a high-resolution (1024 × 768), true-color (24 bits), non-flickering screen (refresh rate of 75 Hz), the amount of information read by the RAMDAC is nearly 170 MB— almost the entire available output for conventional designed DRAM. This leaves us no time for the CPU to load the data to be displayed, or leaves us without the ability to increase the resolution, or to increase the number of colors displayed. VRAM memory is essentially DRAM with a second port added, allowing simultaneous access by both the CPU and the RAMDAC. This dual-ported memory is frequently found in high-end video cards.

WRAM

This stands for Windows RAM. Windows RAM is VRAM that has been modified with two Windows-specific graphic functions. It adds dual-color block write, which helps the video card perform very fast pattern and text fills. It also provides native support for aligned BitBlts (Bit Block Transfers) to perform very fast buffer-to-buffer transfers for full-motion video and 3-D animation by moving blocks of data within the VRAM chip itself, not requiring any CPU intervention.

Shadow RAM

This is a performance technique that copies expansion card and system BIOS ROM into UMB (Upper Memory Blocks). Many expansion cards come with ROM chips containing program subroutines to control the card. Since access to memory locations on a card is limited by the slow speed of the ISA system bus, and by the 16-bit data path of the system bus, overall performance of the card suffers. If you have a 386+ system, the memory mapping capabilities can be used to copy the contents of the ROM chip into fast-access RAM.

If your computer's BIOS supports shadow ROM, you can indicate which ROM address sections are to be shadowed. In addition, the line "shadow option" controls the shadowing of the ROM BIOS routines and video ROM. Some devices can be shadowed using a program supplied by the device's manufacturer. Refer to the specific device's manual for details.

Memory Cache

Caching is the single most important performance enhancement concept, and is a technique used to speed up various parts of the computer. Although it can take many forms, the principle is simple: the cache keeps a copy of the most recently used data and retains it for future use. This is possible because most of what a computer does is very predictable. It operates on a limited number of memory locations at any given time; so for that period the CPU uses a relatively small amount of memory.

To take advantage of this situation, system designers have developed a sophisticated way to place the data needed at any given time in a small amount of very high-speed memory, the memory cache, while the rest of the data is kept in a larger amount of less expensive memory. If a computer normally used 10 percent of a program, it would need to keep just that 10 percent in the cache. When the cache is filled, those locations that are the least recently used (the oldest data) are overwritten with new data.

In a well-designed system, the necessary data can be in the cache more than 90 percent of the time, even if the cache size is only a small fraction of the total memory. Caching strikes a balance between the cost of a small amount of fast SRAM, and a large amount of slower (and less costly) DRAM. This results in near-SRAM performance for near-DRAM prices.

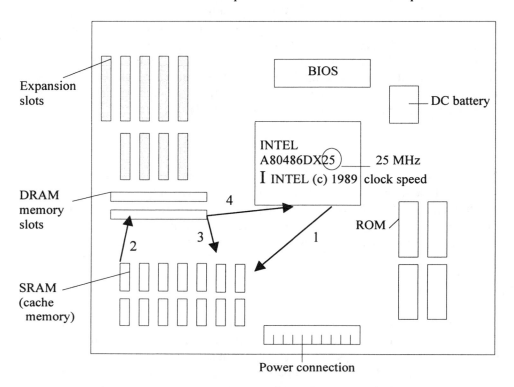

Figure 2.23 Cache memory operation

In more recent CPUs such as the 486, Pentium, Pentium Pro, Pentium II, and Pentium III, an additional layer of cache memory is held inside the CPU chip itself. This means that we a two-level caching system level 1 (L1) is the internal cache within the CPU, and level 2 (L2) is the SRAM cache on the motherboard. Cache memory operates in the following way (see Figure 2.23):

1) The CPU requests data from memory locations (1).
2) The cache controller checks the cache; if it is not found in the SRAM, a request to DRAM main memory is made. (If found, the data is quickly given to the CPU)
3) When the information comes back from DRAM, the cache controller stores a copy in SRAM (for the next time), and forwards the information to the CPU.
4) If the CPU requests the same memory location again, the cache controller gets it quickly from the SRAM, saving time.

If a cached CPU is used, a level 1 (internal) cache search is performed prior to looking in the level 2 (external) cache.

Floppy Drives and Disks

Floppy Drives

Floppy drives are used to allow you to read and write information on floppy disks. PCs usually have one or more drives in either 3.5" or 5.25" formats. When installing a floppy drive, the cabling is most important. The power cable is keyed and you only have to worry about one end. It goes right into the drive socket. The floppy drive cable normally has three connectors and a special twist in a group of five cables to reverse the drive select and motor control signals in the cable. Drive A is attached to the end of the cable, where the wire twist takes place. Drive B is connected in the middle of the cable and the third connector on the other end goes to the floppy drive controller. Floppy drives have a drive select switch or jumper to make them either drive A or drive B. The switch or jumper is normally located near the edge connector at the back of the drive.

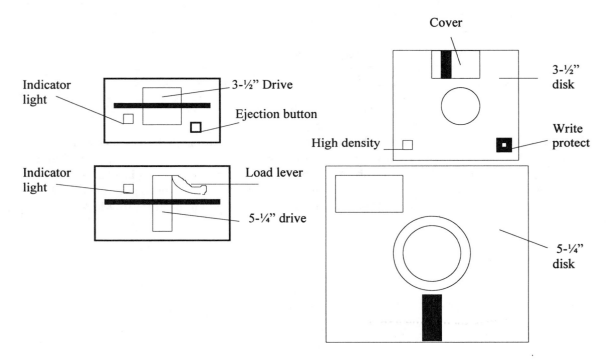

Figure 2.24 Floppy drives and disks

As opposed to the hard disk, the R/W (Read/Write) heads of the floppy do touch the surface of the diskette. Over time tiny pieces of the magnetic oxide flake off and collect on the R/W heads. Other contaminants such as smoke particles, tar from the smoke, and oil from fingerprints accumulate on the heads and cause processing errors and lost data. How often must the heads be cleaned? It depends on the frequency of usage, and quality of the diskettes. However since all cleaning mechanisms are somewhat abrasive, you should minimize the number of times you clean the heads. The two methods for cleaning a floppy drive head are as follows:

• The heads can be cleaned by dipping a lint-free swab in cleaning solution, and wiping the R/W heads.

- A quicker method and easier way to clean diskette heads is to use a cleaning diskette. A cleaning fluid is applied to a fabric coated diskette, and when the R/W head touches the wet surface the particles are softened only to be wiped clean when the disk starts to rotate

Floppy Disks

These are made from a flat, flexible, magnetic medium. There are four types of floppy disks in use today, two high-density and two low-density. They come in two sizes, namely 5.25" and 3.5". Disks are covered with magnetic material. High-density means that a higher-density material and finer grains were used for packing more data into a smaller area.

DOS has to prepare every new floppy although it has tracks and sectors placed on it at the manufacturer. Table 2.2 shows several types of floppy diskette sizes.

Size	Kilo	Bytes	Tracks	Sectors	Sides
5 ¼"	180K	184,320	40	9	1
5 ¼"	360K	368,640	40	9	2
5 ¼"	1,200K	1,228,800	80	15	2
3 ½"	720K	737,280	80	18	1
3 ½"	1,440K	1,474,560	80	18	2

Table 2.2 Floppy disk size table

DOS magnetically divides the sides of the disk into tracks and sectors. DOS also slices each track into evenly sized pieces. Those pieces are called sectors. Every sector is the same size and holds the same amount of information. You can fit 512 bytes per sector. Interleaving is also performed through low-level formatting. Any damaged sectors are marked not to be used at this time. Most floppy disks are low-level formatted at the factory. DOS allows you to perform low-level formatting using the DEBUG command.

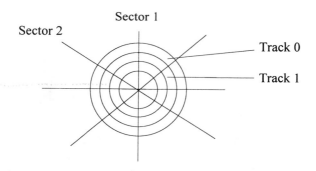

Figure 2.25 Tracks and sectors

In most cases you can easily add a second or new 3-½" floppy drive to your system. Installing a floppy drive is very simple. All you have to do is to follow these steps:

- Power off and unplug the system.
- Remove the cover.
- Remove the plastic cover from the front of the unused bay.
- Insert the new floppy and make sure it's secured.
- Plug the power supply and drive controller into the drive.

- Replace the cover, plug in the system, and cold boot.
- Notify CMOS that you have installed a floppy drive.

If you have two floppies, you can change which drive is A and which is B. You can do this by flip-flopping the cable inside the system and informing the CMOS of the new setup.

Hard Drives

The surface of the disk platter is coated with a layer of magnetized iron oxide particles. The platter itself is made of aluminum and is machined to very demanding specifications. On this surface, the computer stores its data by changing the alignment (magnetizing) these particles. The read/write head is the mechanism that changes the particle alignment when it writes, and can sense the alignment patterns to read back the data.

The R/W head does not make contact with the surface of the platter when the disk is spinning, but glides above the surface on a cushion of air 1/100,000 of an inch thick. If the disk is shaken, if something disturbs the airflow (dirt, dust, smoke), or if an electrical problem causes the disk to slow down, and the R/W head makes contact with the surface, we have a head crash. A crash can scratch the coating off the surface, and the data is obviously lost along with the coating, or the heat caused by the friction can cause the disk to lose its magnetic charge.

While crashes are infrequent occurrences, there are times when the computer is unable to correctly read/write the disk or a few bits of data have been corrupted. DOS will recover by resetting the disk and trying again. DOS will try up to 30 times, and if it is still unable to read/write the disk, will notify you with the famous "Abort, Retry, Fail" message. DOS does not know what to do, and so it turns to you for guidance.

The tracks on the topside of the disk must align with the tracks on the bottom side. The track on side one and the track on side two make up what we call a cylinder. Typically, a 20-MB hard disk drive has 615 tracks. When a group of two or more sectors are combined, they are called a cluster. Most hard disks allocate clusters with two, four, and eight sectors per cluster. When DOS stores a file, it allocates clusters based on the amount of bytes needed by the file. If the file is 10 bytes in size, and DOS is using clusters of two sectors, or 1024 bytes, the other space is wasted. Disk fragmentation occurs when one or more disk files are located in scattered sectors around the disk. Due to disk fragmentation, the read/write head will perform extra revolutions to read all the file information scattered across the various sectors.

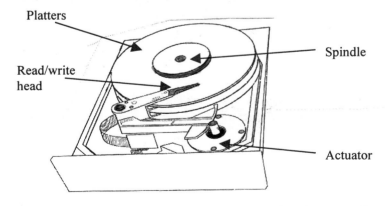

Figure 2.26 Typical internal components of a hard drive

Unlike a floppy drive, data on the hard disk is organized into a combination of tracks and sectors. A hard drive (hard disk) offers storage equivalent to many floppy diskettes. A 10-MB hard disk, for example, equals 28 double-sided floppies. A hard drive is also faster; it retrieves and saves information anywhere from 3 to 15 times faster than a floppy.

Hard disks are usually made of steel platters coated with dense metal oxide on each side. The platters are mounted on a metal spindle. The platter sizes vary from 3 ½" to 5 ¼". Their capacity ranges from 1 to 4 gigabytes. The hard disk has four main parts: the platter, the recording heads, the drive motor, and the drive electronics. The platter spins constantly at 3600 rpm, compared to a floppy, which spins at 300 rpm. You can always calculate the storage capacity of a hard disk using the following equation:

Cylinders × heads

$$SC = Tracks \times Sectors \times Sides \times 512$$

Where:

SC = Storage capacity

Sides = Number of disk sides

Tracks = Number of tracks per side

Sectors = Number of sectors per side

All hard drives and floppy drives should come with a disk controller. The disk controller connects the disk drives, floppy or hard, to the CPU. Without the controller, the CPU would not be able to communicate with the drives, and the computer would have no way to save or retrieve information. The controller card contains the electronics that position the read/write head exactly where it needs to be to write on the disk. The electronics precisely sequence the disk drive activities, translate the encoded bit patterns into actual data, and perform elaborate error checking to tell when things go wrong. All of this is done at extremely high speeds. Encoding data refers to the way the raw data from memory is laid down on the disk surface. So much data is packed into such a small area that the controller would lose track if extra information was not inserted within the data. There are many different hard disk controllers (interface cards) that allow the hard drive or other devices to communicate with the CPU.

FM

This stands for Frequency Modulation encoding. It is a type of modulation where a flux (the process of changing on to off transition) containing a digital bit is going to occur. It is marked by an extra transition called a clock bit. The clock bit forms a periodic train of pulses that enables the system to be synchronized. Two flux changes are needed to record each bit of data; therefore, half the disk space is wasted with this type of encoding.

MFM

This stands for Modified Frequency Modulation. The clock bit is eliminated enabling it to pack twice as much data on a disk as FM encoding. It is a technique used in older disk drives (often low-capacity ST506).

RLL

This stands for Run Length Limited. It translates data into a series of special codes. These codes are chosen for certain numerical properties, particularly the number of consecutive zeros that occur. The logic is extremely complicated, but the outcome is very simple: it packs more data on the disk. RLL packs more sectors on a track, thereby allowing more data to be stored on a disk. By using RLL we can store 50 percent more data in the same space on the disk. Early on RLL had reliability problems, although that is no longer the case. Disk drives designed for MFM encoding cannot be used on an RLL controller.

ST506

This stands for Seagate Technology 506. It transfers data to and from the disk drive in a serial form.
The serial stream of data passes through a deserializer circuit, which converts the data into a parallel form compatible with the computer bus. The speed of the transfer is five megabytes per second. This interface uses two cables: a wide controller cable with 34 pin connections and a smaller data cable with 20 pins.

The disk controller is connected to the drive by two ribbon cables. One has 34 wires and carries the commands from the controller card to the disk drive. If you have two hard drives attached to the same controller, commands will be routed from the controller to the first, and then to the second drive. The second cable contains 20 wires, which carry the data to and from the disk drive. To keep these signals clean, if you have two drives attached to one controller, each drive gets its own 20-wire cable, which you should keep as short as possible.

To eliminate electrical signal reflection, the last drive on the cable must have a terminating resistor block. This usually looks like a small DIP. The resistance will prevent the signals from echoing back down the cable.

ESDI

This stands for Enhanced Small Device Interface. It permits much higher data transfer speeds than other types of controllers. The typical ESDI drive packs twice as much data than other drives onto each track to read during each revolution. Most ESDI drives pack 34 sectors to each track. ESDI uses connections similar to those of ST506, but they are not compatible. The same two-cable system is used.

The major advantage of ESDI is that the clock data separator circuitry is put on the drive itself instead of the controller. Its other important advantage was that since data is recovered from the head signals before entering the cable, their bit rate could be raised without undue degradation. The maximum frequency of bit information transfer for the ESDI design is 24 MHz. Proposed extensions to the ESDI standard were to have raised the maximum frequency to around 50 MHz.

The ESDI interface performs better error checking and is therefore more reliable than ST506, and could support much larger capacity disk drives. ESDI has been replaced by the IDE and SCSI standards, which provide larger capacity and better performance at a lower price.

IDE

This stands for Integrated Drive Electronics. It is unique because the controller is built directly into the disk. The interface therefore requires little electronics and is sometimes incorporated into the motherboard. IDE drives are made to understand PC commands, and so there is no need for the controller to translate. As a result IDE drives are faster in performance than ESDI drives.

IDE drives communicate via a single, straight 40-pin ribbon cable. IDE systems permit two drives to be attached. The drives are differentiated by means of a Master/Slave jumper set on the disk drive. It doesn't matter which connector on the cable you use: the jumper setting determines which is drive 0 and which is drive 1.

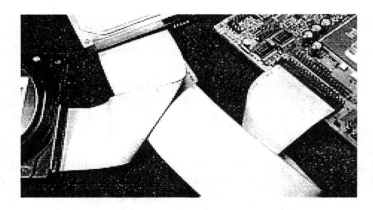

Figure 2.27 Typical two IDE drives using one controller

EIDE

This stands for Enhanced IDE. In an effort to get beyond the limitation of the original IDE/BIOS specifications (e.g. maximum capacity of 528 MB), a new version of the standard was created. EIDE meets this new standard. To implement EIDE you need to upgrade the disk controller, the drive, and the BIOS. EIDE is also designed to be a more universal interface standard, allowing us to attach a wide variety of devices such as CD-ROM and tape drives on the same controller card, as opposed to only hard disk drives with the older IDE standard. It also allows up to 4 devices per controller, as opposed to 2 drives on the IDE controller.

SCSI

(Pronounced _SCUSSY_), this stands for Small Computer System Interface. It is a system-level interface and it provides its own expansion bus to plug into peripherals. SCSI works like a sub-bus: other interfaces plug into it. Up to seven SCSI devices can be daisy-chained to one SCSI port. SCSI uses a parallel connection between the device and the interfaces. This gives a greater potential for more speed. Only a single cable is used to connect the devices.

The SCSI host sends a stream of commands to one of the slaves, which in turn possesses enough intelligence to execute the sequence of operations using the SCSI cable (i.e., it becomes temporary master). When the slave completes its task, it relinquishes the cable, and the original SCSI host may permit some other slave device to control the cable for a while. Thus the disk can be commanded to back itself up on a tape drive without any host computer intervention, allowing it to continue its data processing.

SCSI devices communicate via a single, straight-through 50-wire ribbon cable. The data is carried 8, 16, or 32 bits at a time in parallel. There are relatively few control signals in the cable. The transfer rate is as high as 10 MHz, with a corresponding data rate as high as 40 Mbps using all 32 parallel wires. To differentiate between the drives, each SCSI device must be assigned a Logical Unit Number (LUN). Setting the LUN varies from device to device. Some use a series of jumpers. Others use thumb-wheel switches. Refer to the disk manual drive's manual for specifics of how to set the LUN.

The controller also performs error correction, called the CRC test. The CRC, which stands for Cyclic Redundancy Check, takes a block of data and, through a mathematical formula, yields a long number, which is written just after the data in its sector. When the data is read back, the controller recalculates the number and compares it with the value recorded in the sector. If the numbers don't match, an error occurs.

All hard drives are made of mechanical parts that are prone to fail. When a hard drive fails, it needs to be replaced. There are basically three types of drives in the market today. These drives are ESDI, IDE, and SCSI. The ESDI (Enhanced Small Device Interface) drive connects to a drive controller card while an IDE (Integrated Drive Electronics) drive does not. IDE drives usually have the controller electronics located on the hard drive itself. In most cases IDE drives connect to the motherboard directly. The SCSI drives must connect to a SCSI adapter. Before you purchase any drive, make sure that your system can support it and that you have the appropriate cards needed.

SCSI Adapter

A SCSI adapter allows your system to connect with up to seven high-speed devices. These devices, such as hard drives, CD-ROMs, or tape drives, must be SCSI devices. All SCSI adapters include their own controller electronics onboard. A SCSI adapter creates a second bus to which you can attach devices. Those devices could be internal residing inside your system or external residing outside your system. A SCSI device usually comes with a 50-wire cable terminated with a 50-pin connector. Each device has two ports for incoming and outgoing cables to other devices. When you connect one SCSI device to another, you create a daisy chain, which will expand the bus length. You must use a terminator on the last device's cable port to terminate the daisy chain. This is done to inform the SCSI adapter where the chain ends. Some devices allow you to terminate the device with DIP switches or jumpers. Either way, it must be terminated.

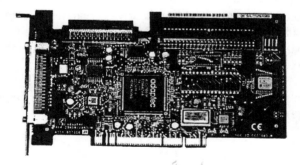

Figure 2.28 Typical SCSI adapter

The SCSI adapter must be powered and priorities assigned by setting the device's SCSI address using DIP switches or jumpers. The higher the SCSI priority address, the higher the device priority will be. To install a SCSI adapter card, you must perform the following steps:

- Use the DOS command MSD or Windows device manager to make sure the card default settings are OK. Otherwise, use the DIP switches or jumpers to change the board settings to avoid IRQ conflicts.
- Power off and unplug the system.
- Remove the cover and ground yourself.

- Remove the expansion card cover and insert the SCSI adapter in the expansion slot.
- Use DIP switches or jumpers to assign the desired SCSI address to each device.
- Cable the devices and terminate the last one.
- Replace the cover.

Two things you should look for when buying a new hard drive are the drive's access time and MTBF. Drive access time is the average time it takes to access data from a hard drive and place it in RAM. The lower access time a drive has, the better it performs. MTBF (Mean Time Between Failure) is a measurement supplied by the manufacturer that gives you a guideline about how long your disk should continue to work before failure occurs.

There are two types of hard drives, internal and external. Internal drives are installed inside the system unit while external drives are connected to a SCSI adapter or to a proprietary card outside the system. Here are the steps on how to install an internal hard drive:

- Power off the system, unplug the power, and remove the system cover.
- Insert the drive into the drive's bay.
- Connect the power cable and data cable to the back of the hard drive.
- Replace the cover, plug the system in, and turn the power on.
- Assign the drive's settings to the CMOS memory. *Write down the drive settings and place them in a safe area. In case the CMOS battery dies, you will need to use them again.*
- Partition the hard drive. This is done to give the hard drive a logical size or divide it into more than one logical drive. You must partition the hard drive to let DOS store information on it.
- Make sure you have a bootable floppy disk with the DOS commands FDISK and FORMAT copied on it.
- To partition the hard drive insert the bootable floppy disk into drive A, and warm boot.
- Type FDISK to partition.
- If you want to partition the hard drive into more than one logical drive, make sure to activate the primary drive before you exit the FDISK menu.
- Finally, you will have to format all the partitioned drives created by using the FORMAT command. For example the command A:\>FORMAT C:/S formats the C drive and makes it bootable and A:\>FORMAT D: formats drive D without making it bootable. You must make the active drive bootable.

After the disk drive has been physically installed in the computer system, three steps must be taken to logically prepare the disk for use by DOS. They are:

- Low-level formatting
- Disk partitioning (FDISK)
- High-level formatting (FORMAT)

Low-level Formatting

This is sometimes called physical formatting. If needed, it is the process of delimiting sector boundaries on the disk surface. New ST506 and ESDI drives come from the factory without any marking, and must be marked before they can be used. Magnetic patterns are made on the disk surface, and the sectors' address and sector delimiters are written, so the disk controller will know when it has arrived at the correct sector. This formatting is not done by the DOS FORMAT command we are familiar with, but rather with a special program built into the controller's ROM, or with a utility program supplied by the manufacturer of the disk controller.

Low-level formatting must be performed once on all new ST506 and ESDI, and some SCSI disk drives. SCSI and IDE drives are low-level formatted in the factory. When a drive is manufactured, its magnetic particles have a random alignment, and the disk controller is unable to determine where a track or sector begins. The low-level formatting process creates the necessary patterns to enable the controller to find a specific sector in the future.

Each sector on the disk has special bit patterns in addition to the data you store. The sector begins with a synchronization pattern to enable the controller's electronics to get ready, followed by the cylinder, side, and sector number that uniquely identifies that sector. The CRCs (Cyclic Redundancy Checks) allow us to determine if any bits have been corrupted.

Hard Disk Partitioning (FDISK)

Hard disk partitioning creates a small table in the very first sector of the disk. This table contains the details of how the disk has been divided. Due to limitations over the years, the maximum size of a hard disk that could be attached to a PC was limited. The physical disk could be subdivided into multiple pieces (partitions) to permit DOS to access all of it. You use the FDISK command to specify the partition sizes.

In DOS versions 2.1 through 3.30, only one partition was available for DOS, and the maximum size was 32 MB. For those who used more than one operating system (e.g., DOS and UNIX on the same computer), a disk could be divided into DOS and non-DOS partitions. DOS can still only access 32 MB, but the other operating systems can access the rest.

With DOS 3.3, the limit of one DOS partition was broken, but each partition can still be no larger than 32 MB, so large disk drives can be handled. Starting with DOS 4.0, the limit of 32 MB per partition was also broken, enabling us to handle extremely large drives up to a maximum of 512 MB. With DOS 5.0 the size was raised to 2 GB.

High-level Formatting

This is sometimes called logical formatting. DOS allows you to perform high-level formatting using the FORMAT command. When high-level formatting is performed, DOS establishes three important areas on the boot sector of the disk. The boot sector area is an area that is right next to the partition table area.

- Boot record—DOS places a boot record on the disk. A boot record is a routine DOS uses to load itself. The boot record consists of the manufacturer's name, version number, number of bytes per disk, number of sectors per cluster, and maximum number of root directory entries. It also consists of the total number of sectors, description of the media, number of sectors per track, number of disk heads, number of hidden sectors and, the bootstrap program.
- File allocation table—DOS sets up two copies of the File Allocation Table (FAT) on the disk. The FAT is responsible for keeping track of where all the files are located on the disk.
- Root directory—DOS sets an area for the beginning of a root directory. The root directory keeps six pieces of information on hand.

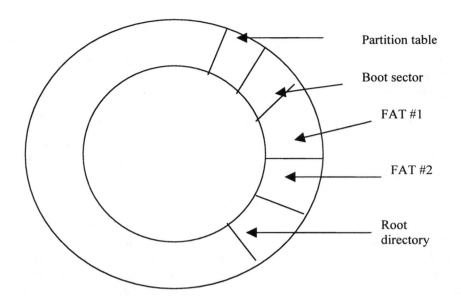

Figure 2.29 Areas on disk after high-level formatting

Modems and Multimedia Sound Cards

Modems

A modem is a hardware device that lets you connect two computers using standard phone lines. First, the sending computer's modem **mo**dulates the computer's digital signals into analog signals that can pass over the phone lines. Next, the receiving computer's modem **dem**odulates the analog signal back into the digital signal that computers understand. There are two types of modems: internal and external. Internal modems are based on a board that can be plugged into any of the expansion slots in the PC. External modems are plugged into the RS232 port or the COM port of the PC with a RS232 cable. A telecommunication software program is needed in order for you to start communication with another computer. The most popular communication programs are PROCOM Plus or PC ANYWHERE. For example, to send a 64-KB (65,536 bytes) file using a 1200-baud modem, you need to send over 524,288 bits (65,536 × 8). A 1200-baud modem sends 1200 bits every second. Therefore to send 64 KB (524,288 bits), the transfer time would take approximately 7 minutes. Table 2.3 shows a list of the approximate time the file transfer would require at different modem speeds:

Modem Speed	File size	Transfer Time
300-baud	64 KB	About 29 minutes
1200-baud	64 KB	About 7.3 minutes
2400-baud	64 KB	About 3.6 minutes
9600-baud	64 KB	About 54 seconds
14.4 K-baud	64 KB	About 36 seconds
28.8 K-baud	64 KB	About 18 seconds
33.6 K-baud	64 KB	About 15 seconds
56 K-baud	64 KB	About 9 seconds

Table 2.3 Transfer times for different modem speeds

With today's success of the Internet, more and more people are installing modems in their PCs to exchange files, chat, send and receive faxes, or send and receive electronic mail (E-mail) messages from other PCs.

Figure 2.30 Typical modem connections

A modem is a piece of hardware that allows you to connect two PCs using a standard telephone line. A fax modem allows you to send faxes to a remote fax machine. Modems come in different speeds. The speed is expressed in bits per second (baud rate). A 9600-baud modem can send or receive 9600 bits of data in one second. Therefore, the faster baud rate your modem has, the less time you will waste waiting for your modem to send or receive information. Modems can be internal or external. Here are the steps to install an internal modem:

- Identify the IRQ and serial port of the modem you will use. Make sure that the IRQ settings are done correctly to avoid conflicts with existing hardware in the system. Also, make sure you use the right COM (serial) port settings to avoid conflicts. Remember that COM1 and COM3 share a common address and COM2 and COM4 also share a common address. For example, if you have your mouse attached to COM1, your modem cannot use COM3. Even though the PC can support four COM ports, only two can be used at the same time.
- Power off, unplug, and remove the system cover.
- Insert the modem card into an open expansion slot.
- Replace the system cover.
- Connect the telephone cable from the wall outlet to the modem port labeled LINE.
- (optional) Connect your telephone to the port labeled PHONE.
- Plug in the power and cold boot.

For fax modems, you need special software that allows you to use the fax, dial, and connect to a remote fax machine. The fax software also lets you receive faxes while you are running other programs. You must also assign data communication settings using your modem software to be able to use your modem. Data communication settings are essential for two computers to synchronize how fast they send or receive information and how much information they will send or receive.

Baud rate is the serial data speed, which is expressed as the number of bits transmitted per seconds (Bps). When transmitting data, one needs a Start bit and a Stop bit. The start is normally 0 and the stop is normally 1. A parity bit is used in transmission of data for error correction. It normally comes before the stop bit. The parity bit can have up to five options, depending on your software. Space parity is a parity bit of zero. An odd parity is a parity bit of 0, meaning there is an odd number of 1s in the transmission, or the odd parity will be a 1 if there is an even number of 1s in the transmission. An even parity is a parity bit of 1, meaning there is an odd number of 1s in the transmission, or the even parity will be 0 if there is an even number of 1s in the transmission. Mark parity is a parity bit that is always 1. No parity is a 0 parity being transmitted.

Synchronous transmission (serial transmission) requires the receiver and a transmitter to be clocked. It is only used exclusively with mainframe PCs. Asynchronous transmission (serial transmission) does not require the receiver and transmitter to be clocked. A block of data could contain 1 start bit, 7 data bits, 1 parity bit, and 1 stop bit. When using asynchronous transmission, you must keep three things in mind. The transmitting and receiving hardware must use the same asynchronous data format. If only text data is being transmitted, it is permissible to use 7 data bits. If binary data is sent, you must use 8 data bits. The best general-purpose asynchronous format is 1 start bit, 8 data bits, 1 stop bit, and no parity.

Simplex communication is a type of data transfer method that requires a communication channel in which information flows in one direction only. An example is a radio or TV station. Duplex communication refers to two-way communication between two systems. Half-duplex communication is a type of data transfer method that requires data to be sent in only one direction at a time. Full duplex communication is a type of data transfer method that requires data to be sent in both directions at the same time. The voltages for sending data are a logic 0 = positive 3 to positive 25 volts and a logic 1 = negative 3 to negative 25 volts; for receiving data, a logic 0 = positive 5 volts and a logic 1 = negative 5 volts.

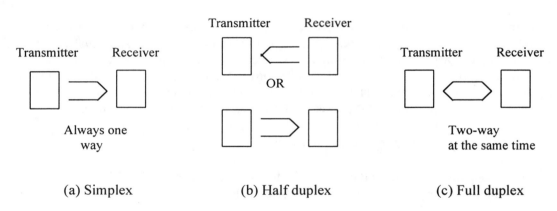

Figure 2.31 Basic communication methods

Multimedia Sound Cards

There are many programs in the market today that use sound cards to make sound effects, audio, video, and playback CD audio. Sound cards can provide recordings via an external microphone, or recordings via a line-in source, MIDI interface, WAV file playback, and audio CD-ROM playback. The cards come in 8, 16, and 32 bits. The higher the bits, the more expensive the sound card will be. This is because it uses more memory, thus providing better sound. You can follow the modem installation steps to install a multimedia sound card. If your sound card or another card stops working while you are using the sound card, you have probably encountered a hardware conflict. Most sound cards will require a specific IRQ setting, an I/O address, and a DMA channel setting. To change the settings, you can use DIP switches and/or jumpers on the board.

Figure 2.32 Typical multimedia card connections

MIDI (Musical Instrument Digital Interface) is a standard used to allow digital electronic instruments, such as an electronic guitar, to interface with a computer and its software. Therefore, you can play back songs and music recorded digitally. You can also use the MIDI port with computer games. You can connect external speakers to the speaker port and you can attach a microphone to the microphone port to record voice or music. The audio-in and audio-out are used to record or send music to a stereo, television, or VCR. You should plug your CD-ROM directly into the audio-in port so you can use the sound card to play back audio CDs.

Monitors and Video Cards

Monitor

In Figure 2.33, a CRT (cathode ray tube) is the picture tube inside the monitor. It consists of an electron gun, a faceplate or a screen onto which phosphors have been deposited, the shadow mask, and the bulb, which holds all these parts together in a vacuum. The monitor itself is sometimes referred to as the CRT.

A high-voltage power supply provides voltage to the CRT with an anode voltage of around 25K. The horizontal and vertical deflector yokes guide an electron gun that projects a fine beam of electrons toward the faceplate. This illuminates the phosphors on the plate. The deflection yoke is made up of two parts: the vertical deflection coil and the horizontal deflection coil. These two coils control the electron beam vertically and horizontally. The horizontal deflection circuit provides the current to the deflection yoke to control the movement of the electron beam in the horizontal direction. The face plate is the part of the CRT that holds the red, green, and blue phosphor dots. They glow when hit by electrons.

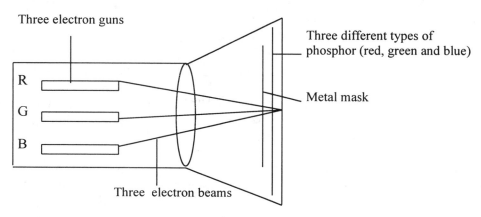

Figure 2.33 A color CRT

The shadow mask is also a part of the CRT. It is a grid that guides the electrons to the faceplate, which ensures that electrons hit only the triads of dots in the cathode. The video amplifier receives three separate signals: red, green, and blue (less than 1 volt in amplitude). The video amplifier amplifies these signals to higher voltages, which are required to drive the CRT cathode. These voltages are then fed to the red, green, and blue cathodes of the CRT. RGB stands for red, green, and blue, the three colors of light that produce any color by varying their intensities.

The angle of the CRT with respect to the horizontal mounting brackets of the chassis is called the tilt. Tilt can vary depending on the monitor's orientation to the Earth's magnetic poles. The method of increasing data densities at conventional scan rates is called interlacing. Half the image is refreshed (every other scan line) to produce a field. Two fields are refreshed at rates of 87 Hz forming one 43.5-Hz frame. In the non-interlacing method, every line of data is drawn on the first pass producing a sharper, flicker-free image. An aspect ratio is the ratio of the width to height of a computer screen. A typical aspect ratio for a monitor is 640/480 = 1.33 or 4 to 3.

Video (television, computer displays) and motion pictures rely on a phenomenon called persistence of vision to give the appearance of smooth motion, in which the brain blends each successive frame seamlessly to create the illusion of motion. If the frames slow down too much, the motion becomes stroboscopic and appears to flicker. Focus is the sharpness of a pixel or series of pixels on the CRT faceplate, also measured as the spot size. The dot pitch is the distance between one phosphor dot and the nearest dot of the same color in the line above or below. Resolution is the number of pixels or dots per linear distance, measured in dots per inch (DPI), and pincushion is an inward bowing of the video image. The pincushion changes per resolution and also according to the size of the image. All monitors experience a slight amount of pincushion.

The screen of a personal computer is divided into a matrix of 80 characters horizontally by 25 characters vertically. Each character requires two bytes of memory storage. Each character is made up of an array of dots. Those dots are sometimes called pixels. There are two types of display systems, TTL and multiscanning. TTL displays (Transistor to Transistor Logic displays) operate on 5 volts as logic 1, and 0 volts as logic 0. Multiscanning displays do not have the horizontal and vertical frequencies synchronized to any standard. They match the synchronous pulses sent from the PC.

Monitors are made to display characters and images by illuminating three different types of phosphors. The monitor display is made up of small groups of phosphors called pixels. When the different types of phosphors are heated, they illuminate red, green, and blue. Each phosphor color is activated by an electron gun that aims at the phosphor and fires to illuminate the color. The electron gun is fired across, then down the screen to display characters or images. To display an entire screen, the gun repeatedly scans the monitor across and down the screen. With each scan, the monitor illuminates red, green, and blue phosphors. To maintain the image, the phosphors must be continually refreshed (reheated). Therefore, the monitor must move the electron gun very quickly at approximately 30,000 times a second for a VGA monitor. The horizontal refresh rate is the measure of every time the monitor starts to refresh a new line of pixels.

Monitor speed is expressed in terms of the vertical refresh rate. Vertical refresh rate is the rate at which the monitor refreshes the entire screen. VGA monitors can have a 60 or 70 Hz vertical refresh rate. This means that the entire screen is refreshed 60 or 70 times a second. Resolution indicates the number of pixels the monitor can display. For example, an VGA monitor with 1024 × 768-pixel resolution is sharper than a VGA with 640 × 480 pixels. Another element that determines the sharpness of the monitor is the dot pitch. Dot pitch is the distance between two successive phosphor colors. Finally, cheaper types of monitors often use the technique of interlacing. Interlacing is when a monitor refreshes every other line as it refreshes the screen. This can produce a wave-like screen appearance that might be distracting. Non-interlacing monitors, on the other hand, refresh the whole monitor line by line, thus avoiding the wave-like appearance. Installing a monitor is very easy. You simply connect the monitor to the video adapter card and then plug in the power on the monitor.

Video Cards

Remember that the computer monitor used by the PC consists of two essential parts, the monitor itself and a video adapter card. The video adapter card interfaces between the motherboard and the monitor itself. It is also important to realize that there are many different monitors and each requires its own special type of video adapter.

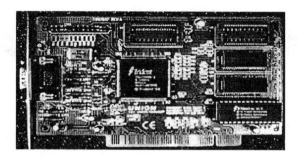

Figure 2.34 Typical video card

MDA

This stands for Monochrome Display Adapter. It has two colors and requires 720 pixels horizontally by 350 pixels vertically. The adapter has a 9-pin connector.

Pin 1 = ground
Pin 2 = ground
Pin 3 = not used
Pin 4 = not used
Pin 5 = not used
Pin 6 = intensity
Pin 7 = video
Pin 8 = horizontal drive
Pin 9 = vertical drive

CGA

This stands for Color Graphics Adapter. Each character on a CGA adapter is made of only 8×8 pixels and only 16 colors are displayed. The resolution is 640×200 pixels. The adapter has also a 9-pin connector.

Pin 1 = ground
Pin 2 = ground
Pin 3 = red
Pin 4 = green

Pin 5 = blue
Pin 6 = intensity
Pin 7 = reserved
Pin 8 = horizontal drive
Pin 9 = vertical drive

HGC

This stands for Hercules Graphics Card. The resolution of this card is 720 × 348 pixels. The HGC technology is now out of date. It had poor resolution, and text as well as graphics was extremely difficult to read.

EGA

This stands for Enhanced Graphics Adapter. The resolution of this card is 640 × 350 pixels. Each character is made of 8 × 14 pixels. It can display up to 64 colors. The adapter also has a 9-pin connector.

Pin 1 = ground
Pin 2 = secondary red
Pin 3 = primary red
Pin 4 = primary green
Pin 5 = primary blue
Pin 6 = secondary green/intensity
Pin 7 = secondary blue / mono video
Pin 8 = horizontal drive
Pin 9 = vertical drive

VGA

This stands for Video Graphics Array. The resolution of this card is 640 × 480 pixels. Voltage levels of the video signal determine the brightness of the image on the screen. The enhanced VGA card is called SVGA (Super VGA). The resolution of this card is 1024 × 800 pixels. The VGA and SVGA adapters have a 15-pin connector.

Pin 1 = red video
Pin 2 = green video
Pin 3 = blue video
Pin 4 = monitor ID bit 2
Pin 5 = ground
Pin 6 = red return
Pin 7 = green return
Pin 8 = blue return
Pin 9 = key blank hole
Pin 10 = sync return
Pin 11 = monitor ID bit 0
Pin 12 = monitor ID bit 1
Pin 13 = horizontal sync
Pin 14 = vertical sync
Pin 15 = not used

The job of a video card is to inform the monitor how to display characters and images on the screen. Most video cards are usually placed in an expansion slot on the motherboard. If the video card resides directly on the motherboard and you need to upgrade it, you must disable the motherboard video using the jumpers or DIP switches on the motherboard. A good video card should have a video accelerator. Video accelerators are chips added on the video card to improve windows and multimedia operations. Video cards differ in speed, the number of colors the card can display, and the card's resolution. Video cards come in 8-bits, 16-bits, 24-bits, and even 32-bits. These represent the number of bits the card uses to represent each pixel's color. Table 2.4 shows the number of colors these video card types can display:

Video card	Number of colors
8-bit	256
16-bit	65,536
24-bit	16,777,216
32-bit	4,294,967,296

Table 2.4 Video cards with number of colors

A good video card can display a large number of colors with high resolution. The key factor in determining how many colors a video card can display at different resolutions depends on the card's onboard memory. Video card memory stores the image displayed on the screen. The more memory a video card contains, the more colors the card can display at one time. For example, if your monitor has a resolution of 640×480, the card must be able to hold values for 307,200 pixels. For a display with 256 colors, each color requires 8 bits (1 byte) of data; thus the card must have at least 307,200 bytes of memory onboard. For a card that can display 65,536 colors, each color requires 2 bytes of memory; thus the memory onboard should be at least 614,400 ($2 \times 307,200$) bytes. As you can see, the higher the resolution, the more memory you should have onboard. Many video cards come with software (their own video BIOS) that can be placed in RAM to accelerate your video output. This software will be a lot faster than the ROM BIOS. By now you should easily be able to install a video card in the expansion slot. Just make sure that you are static free when you remove the system cover.

CD-ROM and Power Supplies

CD-ROM Drive

The amount of information that can fit on one CD-ROM (Compact Disk-Read Only Memory) makes it an excellent candidate for applications that require a lot of storage space, such as animation, detailed graphics, and sound. A CD-ROM drive can be internal or external and you can connect a CD-ROM drive to a sound card or SCSI adapter, depending on the type of drive. Beware of CD-ROM drives that connect to a sound card. If you ever need to change sound cards, you might not be able to use your CD-ROM drive.

When CD-ROM drives were first released, the drives could transmit up to 150,000 bytes (150 KB) of information per second. Today, most drives can transmit up to 1,200,000 bytes per second. These are referred to as 8× (eight speed) drives. The faster the drive, the more information the drive can provide to the computer in a short period of time. The larger the amount of transmitted video data, for example, the more realistic the video's appearance.

The CD-ROM has a detector (called the photodetector) and it is the most important component of the CD-ROM. It contains the electronics necessary to decipher the CD surface into a pattern of bits. Within the detector is an electronic device called a laser diode. This diode emits a laser beam that travels through a series of lenses and other optics to focus the beam on the CD surface. The laser beam reflects back into the detector after hitting the CD surface. The optics within the detector redirects the reflected beam to a photodiode, which detects the amount of light that gets reflected. The CD surface is shiny aluminum protected by a plastic coating. It contains microscopic pits and flat areas called lands. A pit reflects light poorly; a land reflects light well. A weak reflection (pit) indicates a low bit (0), whereas a strong reflection (land) indicates a high bit (1). The CD-ROM drive strings the bits together by spinning the CD to find the successive bits within a sector.

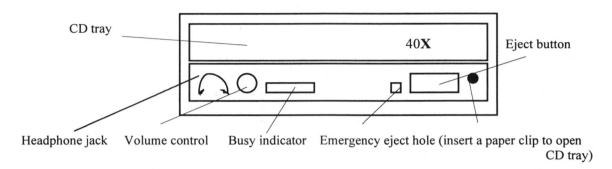

Figure 2.35 The front panel of a typical 8× CD-ROM

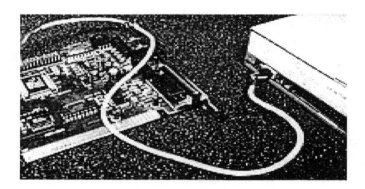

Figure 2.36 Typical CD-ROM with a sound card

Almost every PC you buy today will come with a CD-ROM drive installed. CD-ROM drives are essential to run multimedia-based software. You can connect an internal or external CD-ROM to a sound card, IDE connector, or a SCSI adapter. All CD-ROMs come in different speeds to transfer data to RAM. The faster the CD-ROM speed, the more expensive it will be. The original CD-ROM transferred data at the speed of 150 KB per second (approximately 150,000 bytes per second). Today you can buy CD-ROMs with twice the original speed (2×) or 4×, 8×, 12×, 16×, 32×, and 40×. The larger the amount of transmitted video data, the more realistic the video's appearance will be. However, even the fastest of these drives cannot read data as fast as a hard drive. Table 2.5 shows a list of several CD-ROM speeds:

As long as you have an open expansion slot, you will be able to install an internal CD-ROM. You might want to consider installing an external CD-ROM to have the ability to move it from one PC to another. Always consider connecting a CD-ROM to a SCSI adapter to avoid tying it up with the sound card. If you connect the CD-ROM to a sound card, you may have to replace the CD-

ROM when you decide to upgrade the sound card. Most CD-ROMs come with sound cards, speakers, and a package of multimedia-based software.

Installing an internal CD-ROM drive is very similar to installing a floppy drive. You can follow the steps in installing a floppy drive for the most part. Do not forget that if you are connecting your CD-ROM to a SCSI adapter, you might need to set the drive's SCSI address. Also, depending on the location in the SCSI daisy chain, you might need to terminate the CD-ROM drive. In addition to the physical installation, instead of notifying the CMOS, you will have to install special device driver software before DOS or Windows can access the CD-ROM drive. Normally the installation program will automatically update the AUTOEXEC.BAT and CONFIG.SYS files with the device drivers required to make the CD-ROM operational.

Drive	Speed
1×	150 Kbps
2×	300 Kbps
3×	450 Kbps
4×	600 Kbps
6×	900 Kbps
8×	1200 Kbps
10×	1500 Kbps
12×	1800 Kbps
16×	2400 Kbps
32×	4800 Kbps
40×	6000 Kbps

Table 2.5 CD-ROM speeds

Power Supply

Your power supply regulates the supply of electricity to the various components of the computer. The power supply in IBM PCs ranges from 63.5 to 220 watts. The voltage in the USA is 110 volts at 60 Hz, and in Europe it is 220 volts at 50 Hz. Some computers allow voltages to change depending on the country in which they are used. The power supply provides four distinct voltages. Nearly all the digital circuitry, from the microprocessor to memory, requires 5 volts of DC (direct current). The motors of most disk drives use 12 volts. Serial ports and some other I/O devices require both a positive and a negative 12 volts. A few other components and peripherals also require a negative 5 volts.

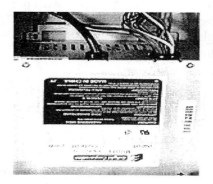

Figure 2.37 Typical power supply and power supply fan

A power supply is engineered to provide up to a maximum amount of electrical power, which is measured in watts. Like an automobile engine, if you exceed its maximum, the power supply is unable to provide you with the amount of energy you need, and will turn itself off. Signs of power supply overload are:

- Random Rebooting—The system reboots itself as if someone has pressed the RESET button on the computer. This is the power supply's way of notifying you that it is having problems. The "power good" line is connected to the CPU's RESET line.
- Overheating—The power supply becomes too hot to touch. Normal operating temperature is slightly above temperature. Overheating can cause what appears to be random system behavior such as locking up, rebooting, random characters on the screen, and crashing. When the internal temperature of the system and its components reaches a high level, the circuitry begins to malfunction, memory chips lose their contents, and the printed circuit boards begin to warp.

By increasing the airflow through the system box and over the expansion cards, we can improve the reliability and increase the longevity of components. To improve the airflow, we can add a fan close to some of the extra openings in the system box. Be sure to replace the metal tabs over any unused expansion slots. This will not only increase the airflow, but also reduce the emitted EMI/RFI (Electromagnetic Interference / Radio Frequency Interference) for the system. Close up any large holes on the bottom or sides of the box. They interfere with the ventilation scheme. The trick is to maximize the airflow over the boards.

PC products are designed to operate in either 110 or 220 volts. If you move your computer from North America (110–120 volts) to anywhere else in the world 220 volts), you must make sure that the power supply can operate on the higher voltage. Look at the power supply: is there a switch to be changed? If yes, make sure it is set properly. If there is no switch, look inside at the specification label on the power supply. Does it say something like INPUT: 100–220V? This will indicate that the power supply is a switched power supply, which will automatically match the power level from the wall outlet. IBM PS/2 and many laptops use this type of power supply. Table 2.6 shows several switch settings and what the results are if the supply voltage is given:

Switch Settings	Supplied Voltage	Results
110	110	Normal operation and setting for the U.S.
110	220	The system will be damaged. The power supply will burn out and need to be replaced.
220	110	No damage, but the system will not operate. Set the selector switch to 110 colts and restart.
220	220	Normal settings for 220 volt countries.

Table 2.6 Switch settings for the power supply

A system board draws about 15 to 25 watts, a floppy drive draws about 10 to 20 watts, a hard drive draws about 10 to 50 watts, and a memory or a multifunction board draws about 5 to 10 watts. The five color leads from the power supply carry different voltages: yellow carries positive 12 volts, red carries positive 5 volts, blue carries negative 12 volts, white carries negative 5 volts, and black is ground. In the fight against power line problems, there are several levels of protection:

- Surge Protectors—These are line filters that clamp high-speed voltage transients.
- Voltage Regulators (line conditioners)—These devices provide more than filtering. They continually filter and regulate the power, providing continuous voltage even during burnout.
- Uninterrupted Power Supply (UPS)—The power supply will be switched over quickly to the UPS battery, within 15 milliseconds, when a blackout occurs. The UPS continuously supplies power from the battery, while at the same time recharging the battery. Backup power can last from a few minutes to a few hours, depending on both the battery's capacity and the computer equipment's power demands. Typically only 15 minutes are needed to gracefully shut the computer down and prevent any disk or file corruption. UPSs are great for riding through a blackout, as its batteries do not run down. When a UPS's batteries run down, it becomes useless. Your system will crash if regular power isn't immediately restored.

Printers

Daisy-wheel Printers

Daisy-wheel printers generate only one font in one size. They are completely obsolete today. See Figure 2.38.

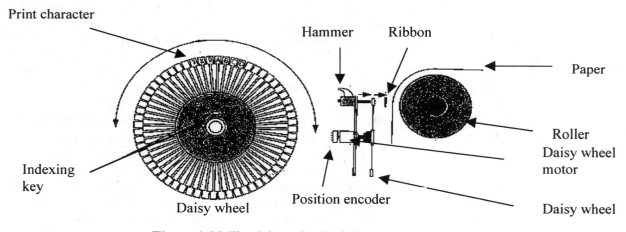

Figure 2.38 The daisy-wheel printer

Dot-matrix Printers

Dot-matrix printers use an array of pins to strike an inked printer ribbon that produce images on paper. You may use either a 9-pin or 24-pin printer. Dot matrix printers treat each page as a raster image. The BIOS for the printer can be either built into the printer itself or as a printer driver. The BIOS interprets the raster image in the same way a monitor does. It "paints" the image dot by dot. Usually the more pins the printer has, the finer the resolution. The disadvantage of a dot-matrix printer is that it needs ongoing maintenance. You must keep the roller, on which the pins impact, and the print head clean with denatured alcohol. There are two types of dot-matrix printers:

1) Impact Dot Matrix (IDM) Printers—An impact dot matrix (IDM) print head system is composed of individual print wires that are assembled into a metal housing carried back

and forth across the page by a carriage system. Each print wire is driven independently by its own solenoid built into the head. Typically print heads contain 7, 9, or 24 separate print wires.

When the host computer sends a character to be printed, a series of vertical dot patterns representing that character (in its selected font and size) are recalled from the printer's permanent memory. The ECP (Electronics Control Package) sends each dot pattern in turn through a series of print wire driver circuits. It is the driver circuits that amplify digital logic signals from main logic into the fast, high-energy pulses required to fire a print wire. As a pulse reaches the firing solenoid, it creates an intense magnetic field that shoots the print wire forward against the page. After the pulse is complete, a spring pulls the print wire back to its rest position.

One of the problems with IDM printing is the eventual building of heat. The substantial current needed to fire a solenoid is mostly given up as heat. Under average use the metal housing dissipates heat quickly enough to prevent problems. Heavy use, however, can cause heat to build faster than it dissipates. Heat buildup happens most often when printing bit-mapped graphics, where many print wires fire continuously. Heating can cause unusual friction and wear in print wires. In extreme cases, uneven thermal expansion of hot pins within the housing can cause pins to jam or bend.

2) Thermal dot matrix print heads—Thermal dot matrix (TDM) print heads replace print wires with individual solid-state dot heaters. Much like IDM print heads, a serial TDM head is assembled as a vertical column of seven or nine dot heaters. The head is mounted on a carriage that carries it back and forth across the page. Every dot can be fired independently by logic pulses amplified with driver circuits. Thermal line-print heads are stationary devices that hold a single horizontal row of dot heaters, one for every possible dot in a row. Because each dot heater in a line consumes relatively little power the driver circuits for a line-print head are usually fabricated into the head assembly itself.

A serial TDM print head works almost identically to an IDM head. When a character is received from the host computer, a series of vertical dot patterns representing that character (including font and size) are recalled from the printer's permanent memory. Each dot pattern is sent out in turn. The dot patterns are then amplified by driver circuits and finally delivered to the print head. Drivers convert the digital information developed in main logic into the high-energy pulses required to fire dot heaters. As a pulse reaches a dot heater, it causes a sudden large temperature rise. This localized heat discolors corresponding points on thermally sensitive paper or melts corresponding points of plastic ink from a thermal ribbon.

Line-point heads are more sophisticated than IDM print heads and depend more on processing power in main logic. After a complete line of characters is sent from the host computer (a line ends when the printer sees a carriage return line feed character), main logic translates the entire line into a series of horizontal rows that span the paper's width. A horizontal row is printed onto thermally sensitive paper (or through a thermal ribbon), the paper advances a fraction to the next adjacent row, and then another row is printed. This process continues until the entire line is complete. If a new row of characters is received, the procedure repeats.

Inkjet Printers / Bubble jet printers

Inkjet printers operate by ejecting ink through tiny tubes by heating the ink with tiny resistors or plates. These resistors are located at the end of each tube. The resistors boil the ink until it creates a tiny air bubble that ejects a droplet of ink onto the paper. The disadvantage of inkjet printers is that the ink tends to dry when not used for a relatively short period of time. To counter this problem, all inkjet printers move the print head to a special position that keeps the ink from drying.

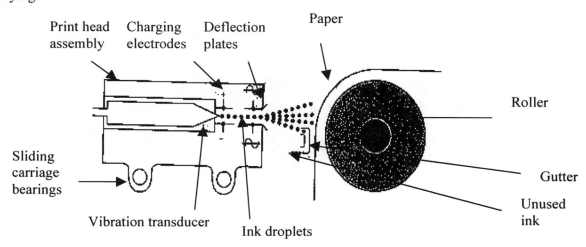

Figure 2.39 The inkjet printer

Non-contact printing is accomplished by spraying liquid ink onto a page from individual ink nozzles. In many ways this method is the same as that used for serial impact and thermal printers., with the notable exception of the use of a liquid ink supply. Depending on the particular print head design, ink is located in a large reservoir away from the head or in a small reservoir integrated into the print head assembly. Inkjet print heads usually hold 9, 12, 24, or more ink nozzles. Each nozzle can be fired independently, so the inkjet system is capable of producing high-quality characters and graphics.

Like other serial print heads, inkjet heads form images one character line at a time as the head sweeps back and forth across the page. Characters that are received from a host computer are converted into a series of vertical dot patterns recalled from the printer's permanent memory. Each dot pattern is sent in turn to the head as it moves. Driver circuits amplify and condition signals produced by main logic into the high-energy pulses that are needed to operate each ink nozzle. As a pulse reaches an ink nozzle, it causes a small piezoelectric ring around the nozzle to constrict. The sudden constriction literally launches a droplet onto the paper. Some inkjet designs replace the piezoelectric ring with heater ring.

Laser Printers

These produce high-quality and high-speed printing of both text and graphics. They are a little more expensive than inkjet printers. Laser printers use the photoconductive properties of certain organic compounds to conduct electricity. Laser printers use lasers as a light source because of their precision. The toner cartridge has several moving parts that suffer the most wear and tear:

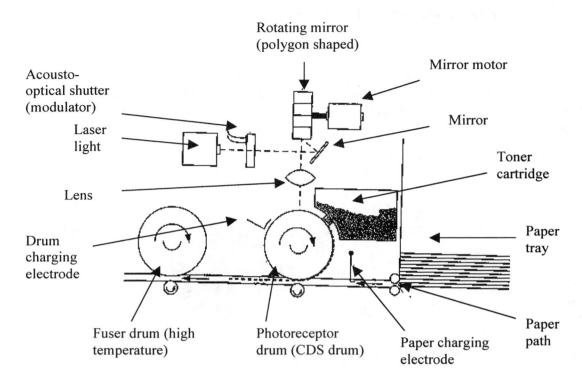

Figure 2.40 The laser printer

- Photosensitive Drum—It is an aluminum cylinder coated with particles of photosensitive compound.
- Erase Lamp—It exposes the entire surface of the photosensitive drum to light, making the photosensitive coating conductive.
- Primary Corona—It is a wire that is located close to the photosensitive drum, but never touches it. When charged with high voltage, an electric field (corona) is formed that allows the voltage to pass to the drum and the photosensitive particles on its surface.
- Toner—It is a fine powder that is made up of plastic particles bonded to iron particles. The toner cylinder charges the toner with a negative charge. The particles of the toner are attracted to the areas of the photosensitive drum that been hit by the laser. The photosensitive drum areas will have a relatively positive charge if they are hit with the laser.
- Transfer Corona—The transfer corona applies a positive charge to the paper, which draws the negatively charged toner particles to the paper. Since the paper itself is charged with positive charge, it is also attracted to the negatively charged drum. To prevent the paper from wrapping around the drum, a static-charge eliminator is placed to remove the charge from the paper.
- Fuser—Since the toner rests on top of the paper after the static charge has been removed from the paper, the toner then must be permanently attached (fused) to the paper. This is done using a pressure roller and a heated roller. The pressure roller presses against the bottom of the page and the heated roller presses down on top of the page, melting the toner into the paper.

Electro-Static (ES) printers must perform two important tasks during initialization. First, a self-test is performed to check the printer's circuits and electromechanical components. This test usually takes no more than 10 seconds from the time power is first applied. Second, the printer's fusing rollers must warm up to a working temperature. Fusing temperature is typically reached within 90 seconds from a cold start. At that point the printer establishes communication with the host computer and stands by to accept data, so its "WARMING UP" code should change to an "ONLINE" or "READY" code.

When a paper tray is inserted, a series of metal or plastic identification (ID) tabs make a set of microswitches. The absence of tabs forms a code unique to that particular paper size. Micro-switches are activated by the presence of tabs, and circuitry interprets this paper type code to know type of media (paper, envelopes, etc.) it is working with.

A mechanical sensing arm detects the presence of paper. When paper is available, this arm rests on the paper. A metal or plastic shaft links the arm to a thin plastic flag. While paper is available, the flag is clear of the paper-out opto-isolator. When the tray becomes empty, however, the arm falls through a slot in the tray. The flag rotates into the paper-out opto-isolator, and the opto-isolator senses that the paper is exhausted. The paper-out opto-isolator is usually mounted on an auxiliary PC board (known as the paper control board).

When paper is available, the paper-sensing arm should move its plastic flag clear of the opto-isolator. When the paper supply is exhausted, the lever should place its flag into the opto-isolator slot. Note: This logic may be reversed depending on the particular logic of the printer. This check confirms that the paper-sensing arm works properly. If you see that the lever mechanism is jammed or bent, repair or replace it.

Printers can be opened in order to perform routine cleaning and EP (Electrostatic Printer) cartridge replacement. The cover(s) that can be opened to access the printer is usually interlocked with the writing mechanism and high-voltage power supply to prevent possible injury from laser light or high voltages when the printer is opened. The top cover (or some other cover assembly) uses a push rod to actuate a simple electrical switch. When the top cover is opened, the interlock switch opens, and the printer's driver voltage (+24 Vdc is shown) is cut off from all other circuits. This action effectively disables the printer's operation. When the top cover is closed again, the interlock switch is reactivated, and printer operation is restored.

An EP assembly uses several tabs (known as sensitivity tabs) to register its presence, as well as to inform the printer about the drum's relative sensitivity level. Main logic regulates the output power of its writing mechanism based on these tab arrangements (i.e., high power, medium power, low power, or no power-no cartridge). Sensitivity tabs are used to actuate microswitches located on a secondary PC board. The sequence of switch contacts forms a code that is interpreted by main logic.

A toner sensor is located within the EP cartridge. Functionally, the sensor is little more than an antenna receiving a signal from the high-voltage ac developer bias. When toner is plentiful, much of the electromagnetic field generated by the presence of high-voltage ac is blocked. As a result the toner sensor only generates a small voltage. An amplifier using some type of operational amplifier circuit often conditions this weak signal that compares sensed voltage to a preset reference voltage. For the comparator, sensed voltage is normally below the reference voltage, and the comparator's output is logic 0. Main logic would interpret this signal as a satisfactory toner supply. As toner volume decreases, more high-voltage energy is picked up by the toner sensor, which in turn develops a higher voltage signal. When toner is too low, the sensed voltage exceeds the reference, and the comparator's output switches to a logic 1. This process is handled in main logic, and a "TONER LOW" warning is produced.

The ES paper transport system is much more sophisticated than the transports used in more conventional serial or line printers. As a result of this additional complexity, main logic circuitry must be able to detect whether or not paper enters and exits the paper path as expected. Several different combinations of conditions can often trigger a "PAPER JAM" condition: paper does not reach the fusing roller assembly within a predefined amount of time; paper does not leave the fusing roller assembly within a predefined amount of time; paper reaches the fusing roller assembly, but fusing roller temperature is below the normal working temperature; finally, if paper is exhausted but the paper sensing arm fails to actuate and trigger a "PAPER OUT" condition, the printer will try to pick up a new sheet of paper. Because paper that is not there cannot possibly reach fusing rollers in any amount of time, this action can also precipitate a "ghost" jam condition. The problem can often be isolated based upon the paper's jam position. Three main trouble areas are the paper-feed area, registration/transfer area, and exit area.

1. Paper-feed area—The paper-feed area consists of the paper tray (and paper), pickup mechanical assembly, and electromechanical clutch. If paper is not reaching the registration rollers, the trouble is probably in this area.

2. Registration transfer area—Registration rollers hold onto the page until its leading edge is aligned with the drum image. Force is supplied by the main motor, but another electromagnetic clutch is employed to switch the registration rollers on and off at the appropriate time. Once the paper and the drum image are properly aligned, main logic sends a clutch control signal. Driver circuits to operate the registration clutch solenoid amplify this signal. After the clutch is engaged, registration rollers carry the page forward to receive the developed toner image. The registration/transfer assembly usually consists of registration rollers, the drive train, a registration clutch solenoid, a transfer guide, and the transfer corona assembly. If paper does not reach the fusing rollers, the fault is probably in the registration/transfer area. You can observe registration roller operation by opening the housing, defeating any corresponding interlocks, and defeating any EP cartridge sensitivity switches, then initiating a self-test.

3. Exit area—At this point a page has been completely developed with a toner powder image. The page must now be compressed between a set of fusing rollers—one roller provides heat, the other pressure. Heat melts the toner powder, while roller pressure forces molten toner permanently into the paper fibers. Fusing fixes the image. As a fixed page leaves the rollers, it tends to stick to the fusing roller. Sets of evenly spaced separation pawls pull away the finished page, which is delivered to the output tray. Main motor force is delivered to the fusing rollers by a geared drive train. There are no clutches involved in exit area operation, so the drive train moves throughout the entire printing cycle.

The scanner is an optical-grade hexagonal mirror driven by a small, brushless dc motor (an induction motor). Printing will only be enabled after the scanner has reached its proper operating speed. The scanner is engaged at the beginning of a printing cycle, so the scanner turns as the main motor turns. For many laser printers, you will recognize the scanner motor by a somewhat distinctive, variable-pitch whirring noise. Motor speed is constantly monitored and controlled by main logic. If the motor fails to turn when power is applied, a scanner "ERROR" is generated.

The fusing assembly is responsible for fixing a toner image onto the page surface. The image results from the application of heat and pressure by a set of rollers. During printing, a high-intensity quartz lamp inside the heater roller establishes a working temperature of about 180 degrees Celsius. The fusing assembly consists of several key components: a fusing lamp, fusing rollers (heat and pressure), a thermistor (temperature sensor), and a thermal switch for emergency over-temperature protection. Power is provided from an ac power supply modulated by a control signal from main logic. This control signal turns the fusing lamp on and off as necessary to develop a stable temperature.

Fusing is an integral part to the successful operation of any ES printer. Toner that is not fused successfully remains a powder or crust that can flake or rub off onto your hands or other pages. Main logic interprets the temperature signal developed by the thermistor and modulates ac power to the fusing lamp. Three conditions will generate a fusing malfunction error: fusing roller temperature falls below 140 degrees Celsius; fusing roller temperature climbs above 230 degrees Celsius; or fusing roller temperature does not reach 165 degrees Celsius in 90 seconds after the printer is powered up. Your particular printer may utilize slightly different temperature and timing parameters. Also note that a fusing error often remains with a printer for 10 minutes or so after it is powered down, so be sure to allow plenty of time for the system to cool.

A thermal switch (sometimes called a thermoprotector) is added in series with the fusing lamp. If a thermistor or main logic failure should allow temperature to climb out of control, the thermal switch will open and break the circuit once it senses temperatures over the preset threshold. This switch protects the printer from severe damage—and possibly from being a fire hazard.

The laser makes a scan line while the main logic waits to send its data. At the beginning of each scan cycle, the laser beam strikes a detector. The detector carries laser light through an optical fiber to a circuit that converts the light into an electronic logic signal compatible with main logic. Main logic interprets the beam detect signal and knows that the scanner mirror is properly aligned to begin a new scan. Main logic then modulates the laser beam on and off corresponding to the presence or absence of dots in the scan time.

Remember that temperature and pressure are two key variables of the EP printing process. Toner must be melted and bonded to a page in order to fix an image permanently. If fusing temperature or roller pressure is too low during the fusing operation, toner might remain in its powder form. The resulting images can be smeared or smudged with a touch.

Static teeth just beyond the transfer corona are used to discharge the paper once toner has been attracted away from the drum. This process helps paper to clear the drum without being attracted to it. An even charge is needed to discharge paper evenly; otherwise, some portions of the page might retain a local charge. As paper moves toward the fusing assembly, any remaining charge forces might shift some toner, resulting in an image that does not smear to the touch but has a smeared or pulled appearance. Examine the static discharge comb once the printer is unplugged and discharged. If any of its teeth are bent or missing, replace the comb.

Questions

1) What must you do before you touch anything inside a PC?
2) What are the physical characteristics of a male and a female port or cable?
3) How do you change a port's gender?
4) What type of port does a serial mouse connect to?
5) What is stored in CMOS memory? Why is it important?
6) Why do you have to write down your system settings and put them in a safe place?
7) What is IRQ 2 generally used for?
8) How can you find the IRQ status of a peripheral?
9) Why is it important not to have two different peripherals share the same IRQ number?
10) How do you change an IRQ setting on a peripheral board?
11) What must you do before you change an IRQ setting on a board?
12) What are the BIOS chips used for?
13) What do SIMM, DIMM, DRAM, and EDO RAM stand for?
14) What must be done after the physical installation of a floppy drive and a hard drive?

15) How many devices can you attach to a SCSI adapter?
16) How do you attach peripherals to a SCSI adapter?
17) What are the two things to look for in a hard drive?
18) What do you do after you have finished the physical installation of a hard drive?
19) List the steps to install a fax modem.
20) What is a MIDI port used for?
21) What is the definition of vertical refresh rate?
22) What is the difference between interlaced and non-interlaced monitors?
23) What does a video card do?
24) Why is it important to have high memory on a video card?
25) What is the difference between an 8-bit and a 16-bit video card?
26) What is the disadvantage of connecting the CD-ROM to a sound card?
27) What is the difference between high- and low-density disks?
28) How do you write protect a 3-½" disk?
29) What is the difference between high-level and low-level formatting?
30) What are tracks, sectors, clusters, and cylinders?
31) Explain the three areas established on a disk when high-level formatting is performed.
32) How does a hard drive differ from a floppy disk?
33) What is the use of a disk controller?
34) What are the advantages and disadvantages of IDE and SCSI drive controllers?
35) What do the following acronyms stand for?

a)	VGA	j)	EISA
b)	SCSI	k)	VESA
c)	HGC	l)	FM
d)	MCA	m)	RLL
e)	ST506	n)	PCMCIA
f)	ESDI	o)	CRT
g)	ISA	p)	RGB
h)	PCI	q)	CGA
i)	EGA	r)	MDA

36) What is the capacity of a floppy drive if it has 2 sides, 62 sectors, and 5355 tracks?
37) How many bits of data can flow through ISA and EISA controller cards?
38) What is the disadvantage of an MCA board?
39) Name four types of buses a motherboard uses.
40) What is the difference between the 8088 CPU and the 8086 CPU?
41) Explain each of the following terms:

    a) Resolution
    b) Interlaced
    c) Pixel
    d) Pincushion
    e) Full duplex
    f) Asynchronous transmission
    g) Synchronous transmission

42) What is the maximum amount of power a floppy drives and a hard drives can draw?
43) What is the difference between a parallel and a serial port?
44) How many serial ports can DOS use at one time?
45) What are the two types of mice interfaces?
46) Where does the word "modem" come from? How does a modem work?

47) How long would it take to transfer a 64-K file using a 28.8-K baud modem?

48) What is the difference between half and full duplex?

49) What are the advantages of using a CD-ROM?

50) Using which type of cable does data travel over a longer distance (serial or parallel)?

51) What IRQ number do COM1 and COM3 have?

52) What does MMX stand for?

53) What does flash BIOS mean?

54) What is the difference between 30-pin and 70-pin SIMM chips?

55) How fast does a 32× CD-ROM drive access data?

56) How many ports can a PC operate at the same time?

57) What IRQ number is COM3 generally assigned to?

58) What does MMX stand for?

59) What is EEPROM?

60) What powers the BIOS?

61) What does DMA stand for?

62) Which type of memory uses transistors in its design?

63) Which cable has a maximum length of 10 feet?

64) What does EDO stand for?

65) What does MTBF stand for?

66) Give three reasons why you want to FDISK the hard drive.

67) What is a standard to allow digital electronic instruments to interface with a computer and its software?

68) Give two signs that the hard drive is overloaded.

69) If the voltage supply is 110 and the switch was set at 220, what would happen?

70) What type of RAM has a 32-bit data width and takes less space on the motherboard?

3

Windows 95/98/2000 Command Lines

In this chapter, you will learn how to use Windows 95/98/2000 command lines to be able to operate, maintain, troubleshoot, and back up your PC. There are many occasions when a computer technician needs to use command lines. For example, when a user is installing a new hard drive, only command lines can be used to partition or format the hard drive. In Windows 95/98/2000, there are four categories of line commands: system commands, directory commands, file commands, and disk commands. Remember, the internal commands are part of the file COMMAND.COM, but the external commands are located on your disk in a subdirectory called COMMAND under the WINDOWS directory. Command lines are not case sensitive—it does not matter if you use capital or lowercase letters.

System Commands

The prompt on most systems indicates the current default drive. When the boot process is finished and the operating system is loaded, you will see the system prompt displayed on your screen. Depending on which drive (C or A) the PC was booted from, you will see one of the following: A:\>_ or C:\>_. To switch from one drive to another, you must type in the drive letter followed by a colon. For instance, if you are located at the C drive and want to switch to the A drive, then you should type A:. To help you manage your system better, commands are issued at the prompt to tell the CPU what tasks to perform. Some of the system commands are as follows:

?   This is an internal command to activate the online help. The online help contains references on all the line commands.

            To execute:

            Type   Command /?   <Enter>

DATE   This is an internal command. Windows always keeps track of the date and time of every file you create or change. The date and time are listed next to the filename in the directory. The command Date will display the date and prompt you to change it.

            To execute:

            Type   **DATE**   <Enter>

TIME   This is an internal command to display and prompt you to change the internal clock time. This is the time DOS uses to keep track of when a file has been created or changed.

To execute:

Type   **TIME**   <Enter>

CLS   This is an internal command to clear the screen and show only the prompt and cursor on the top left-hand side of the screen.

To execute:

Type   **CLS**   <Enter>

VER   This is an internal command to display the version of DOS.

To execute:

Type   **VER**   <Enter>

PROMPT   This is an internal command to change the appearance of the prompt. You can customize the prompt to display anything you want.

To execute:

Type   **PROMPT** *Parameter*   <Enter>

The word *Parameter* should be replaced with one of the following to customize the prompt display.

Parameter	Description
$Q	= (equal sign)
$$	$ (dollar sign)
$T	Current time
$D	Current date
$P	Current drive or path
$V	MS-DOS version
$N	Current drive
$G	> (greater than sign)
$L	< (less than sign)
$B	| (pipe)
$_	ENTER LINEFEED

Therefore, if you are located at the C drive and want to get the prompt C:\>, you should execute the command **PROMPT PG.**

MEM   This is an external command to display the amount of memory used and free memory available on the computer.

To execute:

Type   **MEM**   <Enter>

SYS   This is an external command to make a disk bootable. Making a disk bootable means transferring the files IO.SYS, MSDOS.SYS, and COMMAND.COM to the disk.

To execute:

Type **SYS** A: <Enter>

File Commands

Copying a file will create a duplicate of the original content of that file, but does not remove the original content of the file. For example, if you want to work on a document at home, you can copy it from your computer at work to a floppy disk and then take the floppy disk home. You can also rename a file and delete it. If you do not have very much disk space, deleting files you no longer use is essential.

COPY   This is an internal command to copy a file or files from one drive to another or from one directory to another.

To execute:

Type **COPY** *Filename.ext* A: <Enter>

As a result, the file *Filename.ext* will be copied to drive A. In addition, wildcard characters are used to copy a group of files. For example, to list files that have something in common, one or more wildcards can be used to list specific groups of files.

A **wildcard** is a character that can represent one or more characters in a filename. The " * " character represents one or more characters in a group of files. The " ? " represents a single character in a group of files. The wildcard characters are used to replace part or the whole filename or its extension.

Command	Result
C:\>COPY *.EXE A:	All the files that have the extension .EXE will be copied to drive A.
C:\>COPY A: *.* C:\TEMP	All the files on drive A will be copied to sub-directory TEMP in drive C.
C:\>COPY T?T.* \WP51	All the files that have 3-letter filenames, regardless of what the second character and their extension are, will be copied to subdirectory WP51 in drive C. The files must begin and end with the letter T.

Here are some examples of wildcards using the DIR command. DIR is an internal command to view the content of a directory.

Command	Result
DIR *.EXE	List all the files that have the extension .EXE.
DIR TEST.*	List all the files that have the filename TEST regardless of what the extension is.
DIR K*.*	List all the files that start with the letter K.

DIR ???.*   List all the files that have a 3-letter filename with any extension.

Please note, when using wild cards in a filename, MS-DOS will ignore letters that come after the asterisk (*) up to the period. If the asterisk is used in the extension, MS-DOS will ignore any letters that appear afterwards.

XCOPY   This is an external command to copy files and subdirectories at the same time.

To execute:

Type   **XCOPY** A:*.* C:\CLASS/S <Enter>

As a result, all the files and subdirectories in drive A will be copied to subdirectory CLASS in drive C. Without the switch /S, the XCOPY command acts exactly like a COPY command.

REN   This is an internal command to rename a file.

To execute:

Type   **REN** *Hardware.doc Software.doc*   <Enter>

As a result, the file name *Hardware.doc* will be replaced with the name *Software.doc*. You can also use the wildcard characters with the REN command. For example:

To execute:

Type   **REN**   TEST.* EXAM.*   <Enter>

As a result, all the files that have the filename TEST are renamed with the filename EXAM. You can also make a duplicate copy of a file by renaming and copying the file on the same drive. For example:

To execute:

Type   **COPY** *Hardware.doc Software.doc*   <Enter>

As a result, you will have two files that have the exact same content under two different names, *Hardware.doc* and *Software.doc*.

DEL   This is an internal command to delete files from a disk.

To execute:

Type   **DEL**   *Filename.ext* <Enter>

As a result, the file *Filename.ext* will be deleted. You can also use wildcards with the DEL command.

Command	Result
DEL *.OLD	Delete all the files that have the extension .OLD.
DEL CALL.*	Delete all the CALL files.
DEL F*.DOC	Delete all the files that begin with the letter F and have the extension .DOC.
DEL \CLASS	Delete all the files in the directory CLASS without changing into the directory.

TYPE This is an internal command to display the content of a file on the screen. You can usually view files with the extensions .TXT and .BAT. Some files are in machine code, and you cannot view these. Other files contain a combination of readable and unreadable characters. Files, which you created with a word processing or spreadsheet program, may appear unreadable with the TYPE command, but the program that created them can, read them.

To execute:

Type   **TYPE** *Filename.ext*   <Enter>

EDIT The EDIT command is an external command that initiates the MS-DOS editor. The MS-DOS editor is a simple word processor. It is a text-editing program that is used to create and edit text files. You can use the MS-DOS editor to create and modify system files and batch files, to view text files, and to create and edit short text files. Remember that most word processors can save files as text files. Do not save batch files and other text files in your word processor's file format. The MS-DOS editor cannot interpret them. Most word processors can read the pure text files that you create with the MS-DOS editor.

If you were to type EDIT to invoke the MS-DOS editor, you would see the MS-DOS editor screen. The bar on top of the screen is called the menu bar. The menu bar consists of four pull-down menu options: File, Edit, Search, and Options. Files is used to open, save, and print files as well as start new files and exit the MS-DOS editor. Edit is used to copy and move text. Search is used to find and replace text. Options are used to change the colors of the screen or to locate a file in the help menu. The title bar in the middle of the menu bar shows you the name of the current file. It reads "Untitled" until you save the file and give it a name. The bars on the bottom and right-hand side are called the scroll bars. The scroll bars are used to move the text horizontally and vertically.

Instruction	How to do it
Enter the MS-DOS editor	Type EDIT at the C prompt. Press <Enter>. Press <ESC> to clear the screen. Type in the desired text.
Delete text	Move the cursor in front of the character and press <Delete>, or move the cursor behind the character and press <backspace>.
Delete an entire line	Place the cursor at the beginning of the line and press <Alt> <Y>.
Join two lines together	Place the cursor at the end of the first line and press <Delete>.

Insert a blank line	Place the cursor in the beginning of the line you want to move down and press <Enter>.
To move the cursor	<up arrow> to move one line up
	<down arrow> to move one line down
	<left arrow> to move one character to the left
	<right arrow> to move one character to the right
	<Page up> to move the screen one full page up
	<Page down> to move the screen one full page down.
	<Home> to move to the beginning of the typing line.
	<End> to move to the end of the typing line.
	<Ctrl><Home> to move to the top of the file.
	<Ctrl><End> to move to the end of the file.
To access the system	Press <Alt>, then press the highlighted letter to choose the desired option.
To save and give a file a name	Press <Alt> then <F> for File. Go down to "Save As" and type the name you want. Make sure it only contains up to eight characters and the extension contains three characters only. Then choose OK.
To print	Press <Alt>, File, Print, then choose OK.
To open a file	Press <Alt>, File, Open. Type the desired name or choose the desired name from the list. Use the <Tab> key to move from one option to another. Choose OK.
To exit	Press <Alt>, File, Exit.

COPY CON   This is an internal command to create a text file on the screen. To create the file, you must type the name of the file you want to create.

To execute:

Type   **COPY CON** *Filename.ext*   <Enter>

After you have pressed <Enter>, the cursor will move one line down. Enter your text and press <Enter> after each line. To save and exit a file created by the COPY CON command, you must press <Ctrl> <Z> then hit <Enter>, or press <F6> then hit <Enter>.

MOVE   This is an external command to move a single file or multiple files. If you choose to move a single file, you can rename it.

To execute:

Type   **MOVE** *Source Destination*   <Enter>

Command	Result
MOVE *file.txt* C:\TEMP	Moves the file *file.txt* from the current directory to the subdirectory TEMP.

MOVE *file1.txt,file2.doc* C:\TEMP	Moves the files *file1.txt* and *file2.doc* from the current directory to the sub-directory TEMP.
MOVE *file.txt* C:\TEMP*file1.doc*	Moves the file *file.txt* from the current directory to the subdirectory TEMP and renames the file *file1.doc*.
MOVE C:\TEMP C:\JUNK	Renames the subdirectory TEMP from the current directory to JUNK.

ATTRIB   This is an external command used to assign attributes to DOS files.

To execute:

Type   **ATTRIB** *Parameter drive: pathname* /S <Enter>

The parameters are

+R	=	sets the file to read only.
- R	=	removes the read-only status of a file.
+A	=	sets the archive status.
-A	=	removes the archive status of a file.
+H	=	hides the file from the directory tree.
-H	=	removes the hidden status of a file.
/S	=	directs the ATTRIB to process all the files that reside in a given directory.

DOS automatically labels each file as an archive whenever a file is first created or modified to aid in backing up files from one disk to another and in backing up only those files on the original disk that are either new or recently modified. "Read-only file" means that the content of a file can only be read. It cannot be modified, nor can it be erased from a disk. To hide a system file from the user, you would use the +H ATTRIB parameter. This means a user cannot see the file when he or she types DIR.

Directory Commands

Storing groups of files in different directories makes files easier to find. A directory contains a table of files and subdirectories for a disk. A subdirectory is a directory that is inside another directory. For example, in C:\WINDOWS\SYSTEM, SYSTEM is a subdirectory of the WINDOWS directory.

DIR   This is an internal command to view the content of a directory.

To execute:

Type   **DIR**   <Enter>

Result

Volume in drive C is MS-DOS_6
Volume Serial Number is 1D56-65E2
Directory of C:\

WINDOWS	<DIR>	08-07-95		11:34p
DOS	<DIR>	05-23-93		4:30p
CONFIG	SYS	250	11-11-94	10:04a
AUTOEXEC	BAT	345	12-04-95	5:30p

Note that all the files and subdirectories are in the main or **root** directory of your drive. WINDOWS is a subdirectory of the root directory C. It was created on 08-07-95 at 11:34 PM. The file CONFIG has the extension .SYS and requires 250 bytes of drive space to be stored.

Command	Result
DIR /P	Lists the directory content and pauses when the screen is full.
DIR /W	Lists the directory content horizontally on the screen. Note that only filenames are listed. No other information about the files' size or date and time of creation appears.
DIR /W /P	Lists the directory content horizontally and pauses when the screen is full.
DIR A:	Lists the directory content of drive A.
DIR Filename.ext	Finds the file Filename.ext in the directory you are in
DIR *.BAT /S	Finds all the files that have the extension .BAT anywhere on your disk, including the subdirectories.

MD   This is an internal command to create or "make" a directory. The naming convention is the same as for a filename. The name of a directory cannot have more than eight characters in length and cannot include the characters * . \ / and ?.

To execute:

Type   **MD** *Name*   <Enter>

Name is the name of the directory you want to create.

RD   This is an internal command to "remove" a directory.

To execute:

Type   **RD** *Name*   <Enter>

Remember to remove a directory *Name*, the directory must be clear of files and subdirectories. In addition, the directory must be closed.

PATH   This is an internal command used to tell MS-DOS which directories it should use to find executable files. By default, the current drive and directory the user is in is also part of the path.

To execute:

Type   **PATH** *C:\DOS;C:\WINDOWS* <Enter>

Notice that a semicolon separates each directory. If you want to execute any executable files from the C:\> prompt, DOS will search your current directory and the DOS and WINDOWS directories for that specific file to be executed. You can always display the path statement just by typing the word PATH.

DELTREE   This is an external command used to delete a directory, its files, and subdirectories, and all files within the subdirectories. This is much faster than deleting the files with **DEL** and then removing the directory with **RD**.

To execute:

Type   **DELTREE** */Y Path*   <Enter>

/Y is an optional switch used to suppress prompting for permission to delete. Be very careful when using the */Y* switch. You can easily delete more than you intend.

CD   This is an internal command to open or "change" to a directory.

To execute:

Type   **CD** *Name*   <Enter>

As a result, the prompt C:*Name*> will be displayed.

Command	Result
C:\>CD WINDOWS\TEMP	C:\WINDOWS\TEMP>. You will automatically be located at the TEMP subdirectory of WINDOWS.
C:\WINDOWS\TEMP>CD..	C:\WINDOWS>. You will close the last open subdirectory.
C:\WINDOWS\TEMP>CD\	C:\>. Closes all directories and returns to the root directory.

Disk Commands

DOS has to prepare every new floppy and hard disk before use. Formatting the disks does this. Since different operating systems prepare the disks differently, manufacturers typically sell their disks unformatted.

VOL   This is an internal command to display the disk volume label and serial number.

To execute:

Type   **VOL**   <Enter>

LABEL   This is an internal command to create, change or delete a volume label name of a disk. You can internally label a disk using this command. The name can have up to 11 characters.

To execute:

Type   **LABEL**   <Enter>

FORMAT   This is an external command to format a disk. Formatting a disk will do the following: Erase all the information on the disk, create two copies of the File Allocation Table (FAT), place a boot record on the disk, and place the beginning of a root directory on the disk. The FAT table is responsible for keeping track of all files that are located on the disk. The boot program contains the program bootstrap that initiates the booting process.

To execute:

Type   **FORMAT** A: <Enter>

Remember, if your FAT gets scrambled or the boot record is destroyed, you must reformat the disk. Hence, all data will be lost.

Command	Result
FORMAT A: /S	Formats a disk in drive A and makes it bootable
FORMAT A: /4	
or	
FORMAT A: /f:360	For a 5.25" disk. It formats a low-density disk in a high-density drive
FORMAT A: /n:9 /t:80	
or	
FORMAT A: /f:720	For a 3.25" disk. It formats a low-density disk in a high-density drive

DISKCOPY   This is an external command to make a duplicate copy of a floppy disk. The DISKCOPY command will do the following: Erase the target disk, format the target disk if it was not formatted, and copy all the files, subdirectories, hidden files, volume labels etc. to the target disk. Both disks must be of the same size.

To execute:

Type   **DISKCOPY** A: A: <Enter>, if you have one floppy drive or
Type   **DISKCOPY** A: B: <Enter>, if you have 2 floppy drives

CHKDSK   This is an external command to check the status of a disk. A summary of the disk's usage will be displayed.

To execute:

Type   **CHKDSK**   <Enter>

This command will display how many total bytes the disk has, how many bytes are located in hidden files, how many bytes are in directories and files, and how many bytes are available on disk. The CHKDSK command also provides a brief summary of memory usage.

FDISK   This is an external command used to configure a hard disk for use with DOS. It displays a series of menus to help you partition the hard drive. You must partition the hard drive to give it a logical size or divide it into more than one logical drive. Remember that when you partition the hard drive, all the information on the hard drive will be erased, because you will be required to format it. In addition, you must set one of the partitions to be active for DOS.

To execute:

Type   **FDISK**   <Enter>

You can also use the FDISK command to delete partitions. You must delete the extended partition first, the logical partition second, and the primary partition last. The primary DOS partition is usually the one partition that needs to be activated when it is created.

Batch Files

DOS batch files allow you to group DOS commands in a file. This file must have the extension .BAT. When you type the name of the file with or without the extension, DOS will automatically open the file and execute each DOS command as if it were entered from a keyboard. You may create your batch file using the COPY CON or EDIT commands. For example, the following tasks may be performed one at a time or as a batch file:

a) Go to drive A (**A:**)
b) Delete all the files that have the extension .DOC (**DEL *.DOC**)
c) Go to drive C (**C:**)
d) Open the directory APPS (**CD APPS**)
e) Copy all the files that have the extension .TXT to drive A (**COPY *.TXT A:**)
f) Go to drive A (**A:**)
g) Confirm that the files been copied (**DIR *.TXT**)

You could perform each step a through g interactively, by entering one command at a time, pressing the <Enter> key, waiting for the command to act, and following up with an action of your own. However, a batch file would group steps a through g and execute them automatically. To create a batch file using the internal command COPY CON, do the following:

Type   COPY CON TEST.BAT
Type   A:
Type   DEL *.DOC
Type   C:
Type   CD APPS
Type   COPY *.TXT A:
Type   A:
Type   DIR *.TXT

Press the keys <F6> and <Enter> or <Ctrl> and <Z> to save and exit the file. To execute the file, just type the filename TEST without the extension at the C prompt. Steps a through g will be executed automatically. Now you can execute these commands any time you need them just by typing TEST at the C prompt.

If you wanted to make changes to the batch file using the COPY CON command, you would have to type COPY CON TEST.BAT and retype all the commands all over to insert the changes you want. On the other hand, using the external command EDIT, you do not have to retype all the commands to make changes. You may create a file using the EDIT command just by typing EDIT TEST.BAT. Type all the DOS commands you want, save, and then exit. Execute the batch file that was created using the EDIT command the same way as when using the COPY CON command. Remember, all batch files must have the extension .BAT. You may use the EDIT command to edit a file that was created using COPY CON.

What does the following batch file do?

```
CLS
REM THIS IS A TEST BATCH FILE
REM Created by JOHN DOE
CALL CLEAN
DIR
PAUSE Press any key to continue
@ECHO OFF
REM THE CLEANING IS DONE
```

The above batch file will perform the following: The CLS command will clear the screen and place the prompt at the top left-hand corner. The two REM statements are used for remarks. That is, any information typed after the command REM will not be executed. It will be displayed as a comment or remark. The CALL CLEAN command calls a batch file named CLEAN.BAT and it is executed. The DIR command will list all the files and subdirectories. The PAUSE command will stop the batch file from executing, display a sentence, and wait until you press a key. In this batch file, the file will pause and display the comment "Press any key to continue." After you press any key, the ECHO OFF command is executed. The ECHO command is used to turn off characters being displayed on the screen. The command has the form ECHO ON/OFF/message, where ON turns on screen displays, OFF turns off screen displays, and message is an optional string of characters that will appear on the screen. Note that ECHO is ON by default. In this batch file the REM statement following the ECHO OFF command will not be displayed on the screen.

Questions

1) If you are located in drive C, what is the command to switch to drive A?
2) What is the difference among the COPY, XCOPY, and DISKCOPY commands?
3) What is the difference between the EDIT and COPY CON commands?
4) What command do you use to get technical information about your system?
5) What is the difference between the CD.. and CD\ commands?
6) What is the difference among full, incremental, and differential backups?
7) What does fragmentation do to the hard drive?
8) What command do you use to restore your system after using the MSBACKUP command?
9) What is the difference between the "*" and "?" wildcard characters?

10) What happens to the disk when you use the command FORMAT?
11) What happens to the file when it is deleted?
12) What is the FAT table responsible for?
13) What must you do to replace a FAT table on a disk?
14) What is the difference between the RD and DELTREE commands?
15) What happens to a target disk when you use the command DISKCOPY?
16) What must be done after you partition a hard drive?
17) What do you need to supply DOS with when you use the command UNDELETE?
18) What two commands can you use to move files from one directory to another?
19) Write down the correct commands to execute the following tasks: (Always assume you are working from drive C unless otherwise specified)

   a) Rename the file PROJECT.TXT to EXPER.DOC
   b) Make a duplicate of a floppy disk (Assume you have only one floppy drive)
   c) Create a directory called JUNK
   d) Create a batch file called TEST.BAT using an internal command
   e) Hide a file called CONFIG.SYS
   f) Search for the file AUTOEXEC.BAT
   g) Check how much memory the system has
   h) Check how much space is left on the hard drive
   i) Make a disk in drive A bootable without formatting it
   j) Delete a directory that has files in it, using one command
   k) Delete a directory that has files in it, using several commands
   l) Open a subdirectory called CLASS
   m) List all the files that have the extension .BAT
   n) Delete all the files that begin with the letter K
   o) Copy all the files that have only four characters in their filename to drive A. The filenames must begin with the letter M and end with the letter G. Ignore their extensions
   p) Make a duplicate copy of the file CONFIG.SYS on the same drive C. Name the duplicate CONFIG.OLD
   q) Delete all the files in the subdirectory JUNK without opening it
   r) Format a disk in drive A and make it bootable
   s) Clear the screen
   t) If you are located in drive A, copy all the files and subdirectories to drive C
   u) If you are located in drive A, partition the hard drive
   v) Change the system clock

4

Configuring and Operating Windows 95/98/2000

You can think of Windows 95 as the operating system of DOS 7.0 and Windows 4.0 combined. Windows 95 is a very complex piece of software that is made up of thousands of files. New enhanced features have been added to the user interface, communication, file management, plug-and-play hardware, printing, etc.

Windows Basics

Windows is a special type of applications software that creates a graphical environment that electronically emulates an office. It also ties together the hardware and software resources of the personal computer. Windows is graphical, that is, the screen displays symbols or pictures called icons with which you interact. It also allows the user to work with several different programs at once and to share information between programs. Figure 4.1 shows typical Windows screen elements.

Icon

Start button

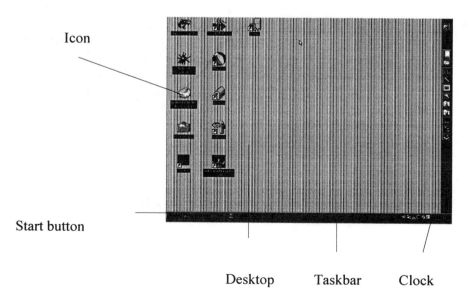

Desktop   Taskbar   Clock

Figure 4.1 Typical Windows 95/98/2000 desktop screen

In Figures 4.1 and 4.2, the Start button used to launch application programs is located on the Taskbar. The Taskbar is used to display which applications are open (active). The Taskbar also displays the clock to show time. The title bar, which is located at the top of each application, is used to identify the application. The menu bar is located below the title bar. It displays the available menu choices. Each application has its own menu. Scroll bars are used to move the screen either vertically or horizontally. The scroll bars are used to view icons that do not show in the window, or text that requires more than the available space. The vertical scroll bar scrolls the screen up and down, while the horizontal scroll bar scrolls the screen left and right.

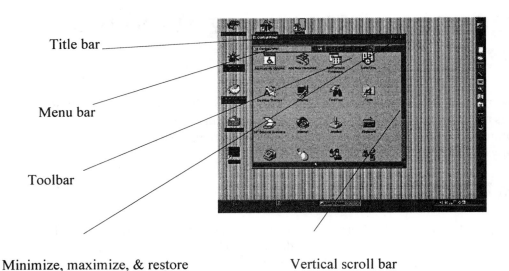

Title bar

Menu bar

Toolbar

Minimize, maximize, & restore   Vertical scroll bar

Figure 4.2 Typical Windows 95/98/2000 screen elements

The Toolbar is a shortcut to icons that you can get from a pull-down menu. For a screen that is not fully displayed on the monitor, you will find the minimize, maximize, and restore buttons located on the top right-hand corner of the title bar. The minimize button is used to shrink the active window to an icon. When the minimize button is pressed, a button icon of the application will be displayed on the Taskbar. The maximize button is used to enlarge the active window to its maximum size. The close button is used to close the active application or window. The restore button, which has the icon,   is used , to restore the active window to its previous size.

Title bar   Menu bar   Minimize   Maximize   Close

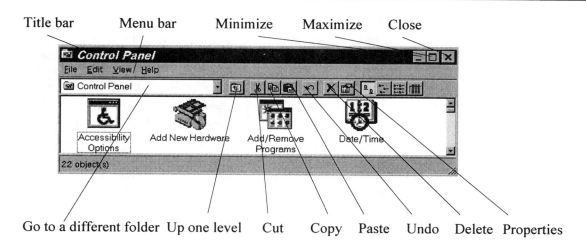

Go to a different folder Up one level Cut Copy Paste Undo Delete Properties

Figure 4.3 A typical Windows 95/98/2000 toolbar

As mentioned before, Windows is a graphical program that takes advantage of using the keyboard and mouse. Moving the mouse requires a smooth, clean surface, and approximately a square foot of space.

Mouse pads are commonly used. It is important for a user to understand mouse terminology. When asked to point, you are required to move the mouse to place the pointer icon onto a menu option, icon, object, or word. When asked to point and click, you are required to point to an object, then press and release the left mouse button. This is used to for selecting and positioning. When asked to drag, you are required to point to an object, hold down the left mouse button, and roll the mouse in the desired direction and then release. When asked to double-click, you are required to point on the object, then use the left mouse button to click twice in rapid succession. This is used to activate and execute.

You can move a window to any place on the desktop. To do so, place the mouse onto the title bar. Press the left mouse button and hold. Drag the window to any desired place on the desktop, then release. You can also change the size of a window by adjusting its width and length. If you were to move the mouse pointer to the edge of an open window, you would notice that the mouse arrow pointer has changed to a sizing pointer. When this happens, just click on the left mouse button, hold and drag the side to any desired length.

Windows 95/98/2000 Fundamentals

Windows 95/98/2000 is a 32-bit operating system. This means that more data can go in and out of the CPU compared to DOS 6.22, which is a 16-bit operating system. Windows 95/98/2000 boots up much like DOS 6.22 does. First, the POST is executed. Second, the two hidden files IO.SYS and MSDOS are transferred to memory. Third, Windows 95/98/2000 will search for the file CONFIG.SYS and load it into memory if found. Fourth, the operating system interpreter COMMAND.COM is loaded into memory. Finally, the operating system will search for the file AUTOEXEC.BAT, and load it into memory if found.

Many new features were added in Windows 95/98/2000, but one of the major enhancements is its hardware plug-and-play option. Much of the hardware currently in your computer is automatically detected upon installation of Windows 95/98/2000. When you install plug-and-play-enabled hardware, Windows 95/98/2000 will automatically assign available interrupts, ports, or addresses without user input.

In Windows 95/98, you may insert commands in the file MSDOS.SYS to tell Windows 95/98 not to execute the CONFIG.SYS and AUTOEXEC.BAT files. You may also edit the MSDOS.SYS file to determine whether you will start in DOS 7.0/8.0 or Windows 95/98 (Windows 4.0/5.0). If you press the key <F8> when the message "Starting Windows 95/98" appears on the screen during boot up, you will be prompted with a startup menu that gives you a set of choices regarding how the computer should start before AUTOEXEC.BAT and CONFIG.SYS are executed. Before you install Windows 95/98, you must know the minimal requirement necessary to run Windows 95/98. Table 4.1 shows the hardware requirements for installing and running Windows 95 only.

COMPONENT	REQUIREMENT
Processor	386 or better.
Memory	4 MB for the operating system. 8 MB required to run most applications. 16 MB recommended for reasonable speed
Video	VGA minimum. SVGA recommended
Disks	3 ½" floppy drive if installing from floppies. CD-ROM , if installing from a CD-ROM.
Hard disk	24 MB to 70 MB of free space depending on how many files you need to install. Leave a minimum of 20 MB of free space for the dynamic swapping during installation.

Table 4.1 Windows 95 hardware requirements

Windows Memory Resources

Microsoft Windows is a program designed to act as an interface between the PC's hardware and other software programs. It provides a standard set of functions, which are used by programs and their programmers. Program developers use the Windows routines to display information on the screen, to create pull-down menus, and to accept the user's responses. With Windows there is a standard for the interface as well for many of the menus. This minimizes the future learning curve for users who have already learned how to use one Windows-based program.

The main Windows resources is memory, and in order to keep Windows running smoothly, you must make sure that you do not run out of it. Using a resource monitoring utility, you can keep an eye on memory usage and be warned when you reach the limits of availability.

The memory resources in Windows are known as heaps. A heap is a programming term for a section of memory reserved for the temporary storage of data. When a program is run, memory is taken as needed from the heap, and when the program ends; it releases the memory for use by another program.

Unfortunately, due to limitations imposed by DOS, there is no way, under Windows 3.x or Windows 95/98/2000, to increase the size of these heaps. With operating systems like OS/2 or Windows NT, these limitations do not exist.

There are a number of different heaps, the ones that need monitoring are listed below. The system and GDI are the ones that need special attention, as they are the ones that are consumed most rapidly.

- System heap—The physical (RAM) and virtual (disk) memory available to hold the programs and data that are currently running.
- GDI heap—This heap holds graphical objects (cursors, fonts, brushes, and pens).
- Menu heap—Holds data relating to menus.
- Text String heap—Holds menu and Windows text strings.
- User heap—Holds all information about open or minimized windows and dialogue boxes.

In the normal course of events, there is enough memory for all these objects. But if you try to load too much, or a faulty neglects to release the heap space when it terminates (a problem known as memory leak), then you will run out of a given resource, which will cause the system to act unpredictably. Either an error message will appear, or the programs will terminate or even cause a system crash.

Microsoft provides a memory resource monitor, which will enable you to see the amounts of heap memory available. The utility must be specifically requested when installing Windows 95/98/2000. Select the custom setup option during the installation process and install all the system tools. Once installed, in order to start the Resource Meter you must click on Start / Programs / Accessories / System tools / Resource meter.

Video Standard for Windows 95/98/2000

One of the fastest growing uses of the PC is animation, full-motion video, and games. Microsoft has developed a series of new standards, and Windows 95/98/2000 has been dramatically enhanced with mechanisms to support them.

- Improved Video Performance—A 60-second video that is 320 × 240 pixels and 15 frames per second requires over 200 MB of video data per minute. That is over 3 MB per second. Even with video compression, the performance of the CD-ROM and video devices can be a bottleneck to a satisfying video experience because of the sheer amount of information involved. Thus, previous incarnations of video for Windows were stressed when playing videos. The perceived symptom was a "jerky" video as opposed to one that smoothly displayed frame after frame.

The performance improvements incorporated into Windows 95/98/2000 make it possible to play high-quality videos in a larger window, potentially up to 640 × 480 pixels with the right hardware. Windows 95/98/2000 includes these performance enhancements:

a) DCI—The Display Control Interface greatly accelerates the rate at which video memory is updated. DCI enables video software to get closer to the hardware.

b) CDFS—The CD File System improves throughput from the CD-ROM by providing an optimized, 32-bit, and protected-mode file system. The improved performance of CD-ROM drives in Windows means you are not waiting on your system while reading data from the drive. Video looks better.

c) Better multitasking—Windows preemptive multitasking minimizes the pauses and delays during video playback. The video continues to play while other processes, such as decompressing additional video from the CD-ROM drive, are running in the background.

- Faster Video—Very few multimedia games have been published for Windows and with good reason: Performance was too slow for graphically intense games such as DOOM. Many games learned to rely on painting directly to the video device and using device-dependent features. Both traits are a no-no for well-behaved Windows applications. Thus, games were relegated to run in DOS. In addition, the GDI and windowing environment was just too slow to support gaming software.

  Windows 95/98/2000 provides a significant improvement that opens the door to game developers. WinG (Pronounced Win Gee) provides for virtually direct access to the display device while remaining compatible with GDI. The compatibility with the existing GUI means that programmers can easily port applications to take advantage of WinG's speed. Developing intense multimedia games for Windows is now practical with WinG. Having a place to play, you can count on vendors to provide the games.

- Display Control Interface (DCI)—DCI is the result of a joint effort between Microsoft and Intel to produce a display driver interface that enables fast, direct access to the video frame buffer in Windows. In addition, DCI enables games and video to take advantage of special hardware support in video devices that improves the performance and quality of video. For example, a video player can take advantage of color-space conversion that enables color conversion to RGB (Red Green Blue) to occur in hardware rather than software. Although DCI enables direct access to the frame buffer, it remains compatible with GUI. DCI is available for all sorts of Windows display devices. Some of the possible hardware-specific features are:

a) Stretching—Stretching enables the hardware to change the size of the image instead of having the software do it. Therefore, the software sends the same number of pixels as before, but the hardware stretches the image to the requested size.

b) Color space conversion—Colors are stored in a video using YUV (broadcast signals that define luminance, hue, and saturation), a method similar to human visual perception. Before the image can be moved to the screen, it must be converted to RGB values. DCI hardware conversion saves a lot of time, potentially up to 30 percent, by freeing the software from this task.

c) Double-buffering—This is the process of displaying the screen currently in the frame buffer while painting the next screen in memory or an additional hardware buffer. Because the new screen is quickly copied to the frame, video playback and animation appear much smoother.

d) Chroma key—This enables two streams of video to merge. A particular color in one of the streams is allowed to be transparent before they merge. This process is similar to the "blue screens" that weather forecasters use on your local news broadcast or movies use for special effects.

e) Asynchronous drawing—When used in conjunction with double buffering, asynchronous drawing provides for faster screen painting outside the frame buffer.

Applications using the video for windows architecture will notice performance improvements automatically. On a 486 DX/2 66 with local bus video, DCI provides reasonable 640 × 480 video at 15 frames per second. This is full-screen video, but you will perceive it as being very jerky video playback. However, by reducing the resolution to a 320 × 240 pixel (a quarter-screen) window, DCI provides smooth video at 30 frames per second.

- DCI Application Program Interface (APIs)—For programmers, Windows 95/98/2000 provides a set of new interfaces that allow them greater access and control for video, sound, multiuser applications and digital joysticks.

a) DirectDraw—Essentially, the new version of DCI. DirectDraw will let applications access hardware video and display surfaces directly.
b) DirectSound—The direct interface to sound cards. A consistent API for mixing and developing digital (WAV) sound. Windows 95/98/2000 will also support general MIDI.
c) DirectConnect—A set of interfaces for developing multiuser application programs.
d) Digital Joystick—Device-level interface for digital joysticks. Developers can write to one digital joystick standard to support multiple vendors.

Windows 95/98/2000 Major Features

This section gives an overview and guidance about the major features of Windows 95/98/2000. Windows 2000 Professional has additional features, which are covered in Chapter 1.

- **MY COMPUTER**

    a) **HARD DISK:** has SCANDISK, DEFRAG, and BACKUP utilities accessible by a right-click
    b) **FLOPPY DISK:** You can copy files quickly to a floppy using drag-and-drop or right-click to format a floppy disk
    c) **CD-ROM:** Pops up by default every time you slip in a new CD
    d) **SHARED VOLUME:** Represented by a hand which appears on the icon when sharing a drive or folder
    e) **NETWORK DRIVE:** Icon will appear when you map a shared volume to a drive

- **NETWORK NEIGHBORHOOD**

    Network connections via phone lines, serial lines and network cards of every kind are shown here.

    a) **ENTIRE NETWORK:** Shows the workgroup PCs by default

b) **NETWORK FILE SERVERS:** Right-click for file server information or double-click to browse any volumes you have access to

c) **WINDOWS NT FILE SERVERS:** Are organized into large units called "domains." You can log into an entire domain at once. The server maintains the security settings

d) **WORKGROUPS:** Finds who is in each workgroup in your neighborhood

e) **WINDOWS WORKSTATION**: Shows if you were locked out by password protection. You can also double-click to see shared files and printers

- **RECYCLE BIN**

  This is similar to the "trash can" in a Macintosh computer. When you delete a file, it is placed in the recycle bin until the bin is emptied or it hits the maximum capacity you have set.

- **CONTROL PANEL**

  a) **PRINTERS:** Changes the properties of a printer

  b) **DIAL UP NETWORKING:** Accesses a network station using a modem and without having a LAN card

  c) **ACCESSIBILITY OPTIONS:** Assists the physically challenged in keyboard, mouse, and display settings

  d) **ADD NEW HARDWARE:** Automatically detects new hardware or picks it up from a list

  e) **ADD/REMOVE PROGRAMS:** To set up new software

  f) **DATE/TIME:** To pick a time zone and set the internal clock

  g) **FONTS:** To install, remove, sort, or view fonts. You can also double-click to see a preview

  h) **JOYSTICK:** To install or configure game adapters

  i) **MAIL/FAX:** To set account options for delivery of e-mail and faxes

  j) **KEYBOARD:** To set the language and the blink and repeat rates of the cursor

  k) **MICROSOFT MAIL POST OFFICE:** To create and manage a Windows workgroup mailing system

  l) **MODEMS:** To install modem drivers and configure communication options

  m) **MOUSE:** With the options, you can use trails and animated cursors

  n) **MULTIMEDIA:** To view and set all of the sound and video settings for multimedia

  o) **NETWORK:** To add, remove, or configure any network components and services

p) **PASSWORDS:** To access mult user settings and passwords for individuals

q) **PRINTERS:** A shortcut to printers in My Computer

r) **REGIONAL SETTINGS:** To set up currency values, decimals, and date and time styles

s) **SOUNDS:** To change your system and program sounds

t) **SYSTEM:** To view all your system information and advanced settings

- **START MENU**

a) **PROGRAMS:** In Programs you can configure the icons for programs, unlike in Windows 3.1, in which you nest groups inside of groups

b) **DOCUMENTS:** This will keep a list of the 15 most recently used documents. You can launch any of these documents just by double-clicking

c) **SETTINGS:** To open the control panel and printers so you can add a printer or manage a queue. It is also used to manage the Taskbar and Start menu

d) **FIND:** To search for a file or a folder by name, date, or content. The search result can be sorted by name, date or size. Then you can rename, copy, delete, or move

e) **HELP:** To access the online help, and also supports keywords and full text searches

f) **RUN:** To launch a program or open a folder. It keeps a list of the most recently run programs in a drop-down list.

g) **SHUT DOWN:** Will allow you to quit Windows, reboot Windows, or log in as a new user

- **TASKBAR**

a) **POSITIONING:** To position the Taskbar along the side, top, bottom, right, or left of the screen. You can also hide the Taskbar and it will appear when you move the mouse to the edge of the screen

b) **BUTTONS:** All the programs that are running are shown on the Taskbar as buttons. This makes it easier to switch between programs

c) **TOOL TIPS:** If you have many programs opened on the Taskbar, their title will be truncated. You can get a full description of application by just holding the mouse over the application

d) **STATUS INDICATORS:** To get more details and to use controls on system functions such as the fax, printer, modem, or volumes on the Taskbar, simply click, double-click, or right-click

e) **CLOCK:** Always appears at the Taskbar

- **MY BRIEFCASE**

  You can have any files synchronized in two different computers. Just move any two files into the Briefcase and go. When you return, hook the two PCs together and choose update from the menu. Windows 95/98/2000 will update both PCs with the latest files.

- **MICROSOFT EXCHANGE INBOX**

  a) **ADDRESS BOOK:** Stores all of your e-mail addresses and fax numbers
  b) **RETRIEVE:** To obtain any information on a fax-back document or enter a document name
  c) **COVER PAGE EDITOR:** To customize a fax cover page
  d) **FAX VIEWER:** To rotate, crop, or zoom on any faxes
  e) **PERSONAL INFORMATION STORE:** To store all of your mail and faxes for multiple online services. It also supports drag and drop

- **NEW APPLICATIONS**

  a) **DIRECT CABLE CONNECTION:** To hook up two PCs together with a serial or parallel cable for a peer-to-peer net
  b) **HYPERTERMINAL:** A toned down version of Hyper Access. It supports Z-modem and ANSI
  c) **WORDPAD:** Replaces Write and handles text or formatted files. Reads and writes in MS-Word 6.0 format
  d) **PAINT:** Has a better interface than Windows 3.x and and a floating tool bar
  e) **PHONE DIALER:** To speed-dial eight numbers
  f) **BACKUP:** To back up files to a disk or QIC. It does not support SCSI tapes
  g) **DISK DEFRAGMENTER:** To defragment the hard drive while running an application
  h) **SCANDISK:** Checks for lost clusters, cross-linked files, invalid dates, times, and filenames from within Windows
  i) **DRIVESPACE:** A 32-bit protected-mode disk compression driver, limited to 512 MB
  j) **NET WATCHER:** Shows who's connected to what on the network
  k) **SYSTEM MONITOR:** This will show file activity, net activity, and CPU load

How to Perform Tasks in Windows 95/98/2000

This section shows how to perform important tasks in Windows 95/98/2000.

Files

You can almost do all file management tasks in the Windows Explorer.

TASK	STEPS
To select files in consecutive order	Click on first file Hold the <Shift> key down Click on the last file to be selected

TASK	STEPS
To select files in random order	Click on first file Hold the <Ctrl> key down Click on the desired files

TASK	STEPS
To unselect selected files	Hold the <Ctrl> key down Click on the desired files to be unselected

TASK	STEPS
To delete files permanently rather than moving them into the recycle bin	Hold the <Shift> key down when delete is selected

TASK	STEPS
To copy or move files	Select the files to be moved or copied Right-click on the selected files Select copy or cut Open the destination location Right-click on the destination location Paste the selected files

Drives

TASK	STEPS
To check for 32-bit disk access	Go to the "Control Panel" Select "System" & click on the "Performance" folder Notice that File System is 32 bits & Virtual Memory is 32 bits (see Figure 4.4)

TASK	STEPS
To use swap files in Windows 95	Go to the "Control Panel" Select "System" & click on the "Performance" folder Click on "Virtual memory" (strongly recommended). Click on "Let Windows manage my memory settings" (see Figure 4.5)

Figure 4.4 To check for 32-bit disk access

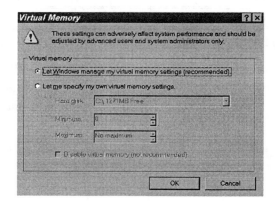

Figure 4.5 To use a swap file in Windows 95/98/2000

TASK	STEPS
To format or copy a floppy disk	Go to "My Computer" Click once on drive A Click on "File" Click on format or copy (see Figure 4.6)

TASK	STEPS
To make Windows 95/98 show files that have the extensions .DLL, .SYS, .VxD, .386 and .DRV	Go to Windows Explorer Click on "View" Click on "options" Click on "Show all files" Click on "Apply" (see Figure 4.7)

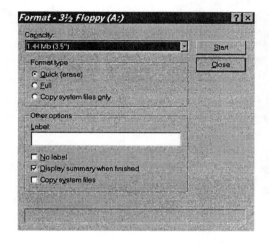

Figure 4.6a To format a floppy disk

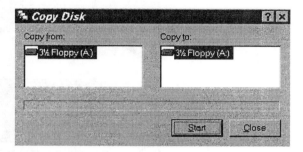

Figure 4.6b To copy a floppy disk

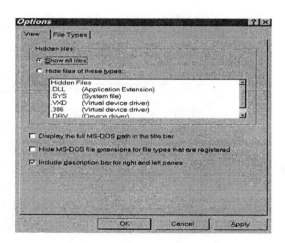

Figure 4.7 To show all files with extensions

Display

TASK	STEPS
To select 256 colors, change the resolution, or check for correct video driver	Go to the "Control Panel" Go to "Display" Click on the "Settings" folder Go to "Color pallete" to select the number of

	colors Go to "Desktop area" to select the different resolutions Go to "Change Display Type" to select video card or monitor type (see Figure 4.8)

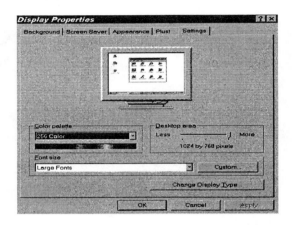

Figure 4.8 To change Windows 95/98/2000 settings

TASK	STEPS
To find the version of Windows 95/98/2000 installed	Go to the "Control Panel" Click on "System." Notice that here the system is "Microsoft Windows 95" (see Figure 4.9)

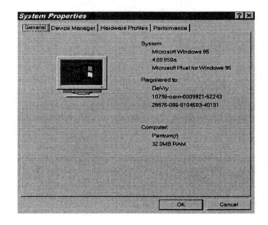

Figure 4.9 To find the version of Windows 95/98/2000

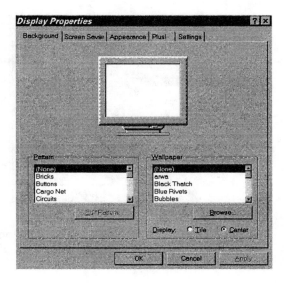

Figure 4.10 To change the wallpaper or screen saver

TASK	STEPS
To change wallpaper or screen saver	Go to the "Control Panel" Go to "Display" Click on the "Background" folder to select different type of wallpaper Click on the "Screen Saver" folder to select a different type of screen saver Click on the "Appearance" folder to customize the desktop style, icon fonts and colors Make sure you click on the "Apply" button for the changes to take effect (see Figure 4.10)

System

TASK	STEPS
To correct date, time, or time zone	Go to the "Control Panel" Click on "time/date" (see Figure 4.11)

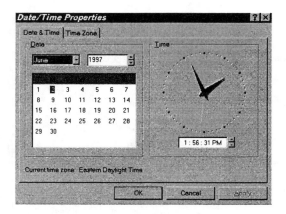

Figure 4.11 To correct date, time or time zone

Mouse

TASK	STEPS
To change the mouse button function, pointers, mouse trails, or mouse speed	Go to the "Control Panel" Click on "Mouse" Click on "Button" for button function, then: Click on button configuration for right-handed or left-handed. Slide the mouse speed knob to adjust click speed Click on "Pointers" for the pointer's function, then: Click on "Normal Select." Double-click on the "icon" pointer. Click on OK Make sure not to select the "icon" pointer from the list, i.e., "Normal Select," "Help Select," or "Working in Background." They represent an actual task and not a mouse pointer. (see Figure 4.12)

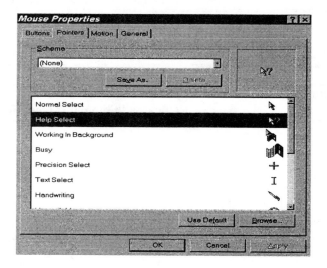

Figure 4.12 To change the mouse pointers

TASK	STEPS
To install all mouse pointers that Windows 95/98/2000 has after Windows "typical" or "normal" installation; does not show all mouse pointers from browse windows	Insert the MS Windows 95/98/2000 CD into the CD-ROM drive Go to the "Control Panel" Double-click on "Add/Remove Programs" Select "Windows setup" Double-click on "Accessories" Click once on "Mouse Pointers" to mark an X Click on "OK," then click on "Apply," then click on "OK" For "Motions," make sure to mark an X in pointer trails (see Figure 4.13)

Multimedia

TASK	STEPS
To open a media player while an AVI file is playing	Double-click on the title bar of the AVI Double-click on the media title bar to exit

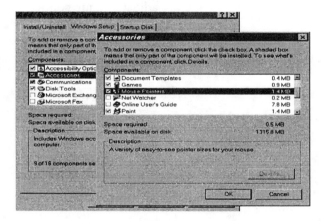

Figure 4.13 To install all mouse pointers

TASK	STEPS
To show the volume control icon next to the clock icon on the Taskbar	Go to the "Control Panel" Double-click on the "Multi-media" icon Select the "Audio" folder Under the "Playback" section, check an "X" next to "Show volume control in Taskbar" (see Figure 4.14)

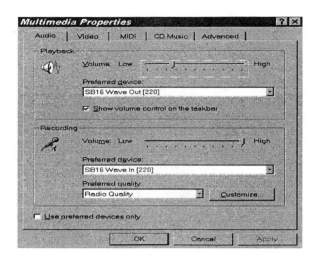

Figure 4.14 To show the volume control icon next to the clock icon on the Taskbar

TASK	STEPS
To use the volume control in Taskbar	Click once on the volume icon to view and adjust volume control Click twice on the volume icon to view and control all volume control options (see Figure 4.15)

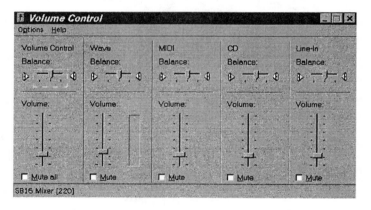

Figure 4.15 To use volume control in the Taskbar

TASK	STEPS
To change the display window when any file is played	Go to the "Control Panel" Go to "Multimedia" Go to "video" Select "window" or "full screen" PLEASE NOTE that full screen may not give the best quality of video. Select window or some other type of video from the list for better view

TASK	STEPS
To make the MIDI player play MIDI files	Go to the "Control Panel" Double-click on the "Multimedia" icon Select the "MIDI" folder Select a single instrument "OPTiMad16 Pro FM MIDI" or select "custom configuration," then: Click on "configure" Make sure that each "channel" is set to *instrument* "OPTiMad16 Pro FM MIDI." If NOT, then: Select the "channel" Click on "change" Click on the down arrow for "instrument" Select "OPTiMad16 Pro FM MIDI" Click on "OK" Go through all 16 channels then click on "OK" and "Apply" (see Figure 4.16)

TASK	STEPS
To remove the "Auto Run" when an audio CD is inserted	Hold the shift key down when the audio CD is inserted, or Go to "My Computer" Click on "View" Click on "Options" Click on the "File Types" folder Double-click on "Audio CD" Click on "Set default" to turn off Auto Run If Play is bold, it plays the CD; if not, it does not play the CD (see Figure 4.17)

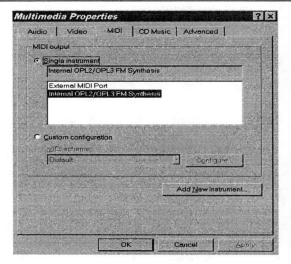

Figure 4.16 To play MIDI files

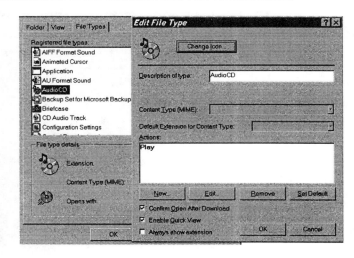

Figure 4.17 To remove the "Auto Run"

TASK	STEPS
To assign sounds to events	Go to the "Control Panel" Double-click on "Sound" (see Figure 4.18)

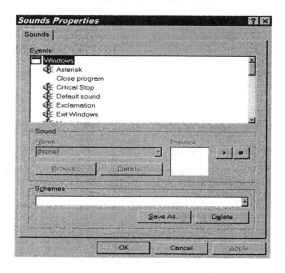

Figure 4.18 To assign sounds to events

TASK	STEPS
To change the cache size for the CD ROM	Go to the "Control Panel" Double-click on "System" Click on the "Performance" Tab Click on "File System" Click on the "CD-ROM" Tab Change the cache size from small to large or vice versa (see Figure 4.19)

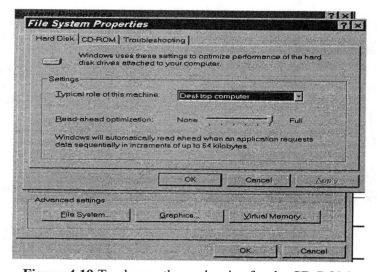

Figure 4.19 To change the cache size for the CD-ROM

My Computer

TASK	STEPS
To rename "My Computer" or any other icon except the recycle bin	Right-click on the "icon" Click on "Rename" Type in the new name

Printers

TASK	STEPS
To install a printer	Go to the "Control Panel" Click on the "Printers" icon Double-click on "Add printer" Follow the instructions on the screen (see Figure 4.20)

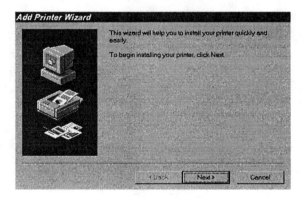

Figure 4.20 To install a printer

Recycle Bin

Remember that the recycle bin icon cannot be renamed.

TASK	STEPS
To free up more space on the hard drive	Check the recycle bin often. Empty the recycle bin by deleting unwanted files.

TASK	STEPS
To recover files that have been deleted	Double-click on the recycle bin icon Select the file or multiple files by holding the <CTRL> key down, then clicking on the desired files to be deleted Click on "File," then click on "Restore"

TASK	STEPS
To set up Windows 95/98/2000 to permanently remove files when they are deleted rather than collecting deleted files in the recycle bin	Right-click on "Recycle Bin" Click on "Properties" Click on "Drive C" Mark an X next to " Do not move files to recycle bin. Remove files immediately on delete" NOTE: You will not be able to recover any deleted files (see Figure 4.21)

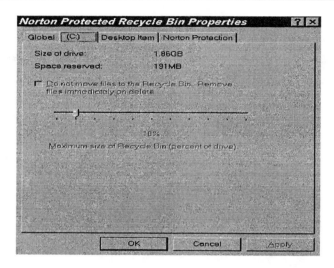

Figure 4.21 To set up Windows 95/98/2000 to permanently remove files

TASK	STEPS
To assign more drive space for the recycle bin to save deleted files	Right-click on "Recycle Bin" Click on "Properties" Click on "Drive C" folder Move the knob to your desired configuration. NOTE: the default is 10% of the hard drive Click on "Apply" (see Figure 4.21)

System Tools

TASK	STEPS
To install disk utility tools and find out how they can be accessed	Click on the "Start" button Go to "Programs" Go to "Accessories" Go to "System Tools" Select what type of disk utility program you want to run (see Figure 4.22)

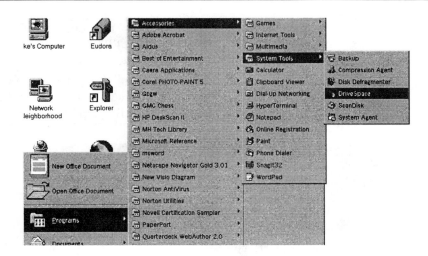

Figure 4.22 To install disk utility tools and find out how they can be accessed

TASK	STEPS
To install the backup utility	Go to the "Control Panel" Double-click on "Add/Remove Programs" icon Click on the "Windows Setup" folder Mark "X" next to "Disk Tools" Click on "Apply" Click on "OK" You should be able to see the "Backup" icon under the "System Tools" which supports both floppy and hard drive (see Figure 4.23)

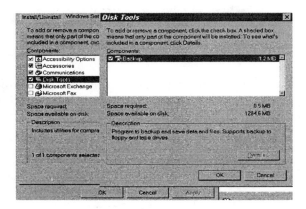

Figure 4.23 To install the backup utility

TASK	STEPS
To back up the entire system	Go to "Programs" Go to "Accessories" Go to "System Tools"

	Go to "Backup," then click on "File"
	Click on "Open File Set"
	Select "Full System Backup"
	Click on "Open," then click on "Next Step"
	Click on "Desired Drive"
	Click on "Start Backup" (see Figure 4.24)

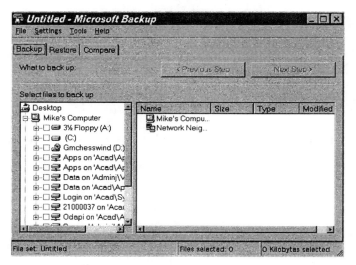

Figure 4.24 To back up the entire system

TASK	STEPS
To format or erase a tape drive	Go to "System Tools"
	Select "Backup," then select "Tools"
	Choose "Format" or "Erase"

Modem

HyperTerminal allows the user to save different customized configuration. It will save telephone numbers, names, icons, dialing properties, communication protocol, etc. To activate the HyperTerminal, the user modem should be configured properly under "Modem" in the "Control Panel"

TASK	STEPS
To use HyperTerminal	Click on "Start" button
	Click on "Programs"
	Click on "Accessories"
	Click on "HyperTerminal"
	Double-click on "hypertrm.exe"
	Type hypterm
	Click on "OK"
	If the icon is not found, then:
	Type in "Name"

	Select "icon" Check the "Dialing Properties" Select "Modify" for modem properties (see Figure 4.25)

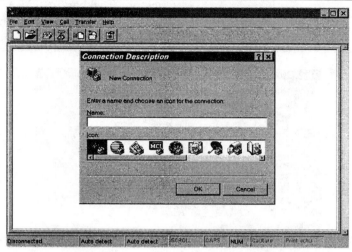

Figure 4.25 To use HyperTerminal

Start Menu

TASK	STEPS
To bring up the start menu	Simultaneously press the keys <Ctrl> and <Esc>

TASK	STEPS
To view and/or modify the start menu content	Right-click on the "Start" button and select "Open" or "Explore"

TASK	STEPS
To add an object (folder, shortcuts, programs, or documents) to the top of the start menu	Drag the object to the start menu

TASK	STEPS
To open the Taskbar content menu when the Taskbar is full	Move the mouse pointer to the edge of the Taskbar and enlarge it

TASK	STEPS
To minimize all windows or open or close a parent folder's window from the Taskbar	Right-click on an empty area on the Taskbar

TASK	STEPS
To empty the document folder from the start menu	Click on "Start" Go to "Settings" Click on "Taskbar" Click on "Start Menu Programs" folder Click on "Clear" (see Figure 4.26)

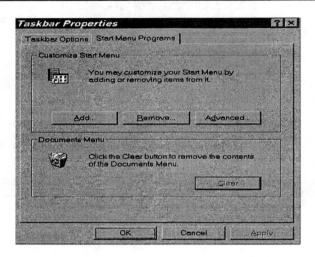

Figure 4.26 To empty the document folder from the start menu

TASK	STEPS
To hide the Taskbar	Click on "Start" Go to "Settings" Click on Taskbar Click on Auto hide and click on "OK" (see Figure 4.27)

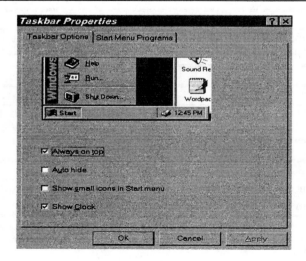

Figure 4.27 To hide the Taskbar

TASK	STEPS
To move the Taskbar to a different position on the screen	Drag the Taskbar by pointing at the Taskbar, holding down the left button of the mouse, and moving to any side of the screen

TASK	STEPS
To add a new folder to a new menu	Click on "Start" Go to "Settings" Click on "Taskbar" Select "Start Menu Programs" Click on "Advanced" Select an appropriate folder where you want to add a new folder Click on "File" Click on "New" Click on "Folder" Type in the "Name" you want for that folder Click "File" again Click "Close" Click on "OK" (see Figure 4.28)

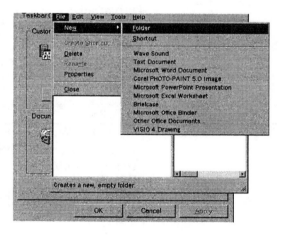

Figure 4.28 To add a new folder to a new menu

Uninstall

TASK	STEPS
To uninstall Windows 95/98/2000	Go to the "Control Panel" Double-click on "Add/Remove Programs" Click on "Windows Setup" Click on "Add/Remove" Follow the instructions on the screen

Boot Options for Windows 95/98

Here is a list of startup options and shortcut keys

STARTUP OPTIONS | DESCRIPTION

STARTUP OPTIONS	DESCRIPTION
<F5>	Starts Windows 95/98 in safe mode, bypassing startup files and loading minimal drivers
<Shift> <F5>	Starts Windows 95/98 in safe mode command line only, bypassing startup files
<Ctrl> <F5>	Runs Windows 95/98 in safe mode, bypassing both startup files and compression drivers
<F4>	Starts the computer using the previous version of MSDOS, if Bootmulti=1 in the Windows 95/98 version of MSDOS.SYS
<F6>	Starts Windows 95/98 in safe mode without networking
<F8>	Displays the Windows 95/98 startup menu
<Shift> <F8>	If the Windows 95/98 startup menu is already on the screen, displays line-by-line confirmation of startup files

TASK	STEPS

At bootup, to be able to select the previous version of DOS	Open the "Windows Explorer"
	Right-click on the "MSDOS.SYS" file at the root directory C:\
	Select "Properties"
	Unmark the "OFF" of the hidden and read-only attributes
	Go to the DOS prompt C:\
	Type edit MSDOS.SYS
	Under "options"
	add the following:
	      bootmenu=1
	      bootmulti=1

Registry

The registry editor is an advanced tool that enables you to change settings in your system registry, which contains information about how your computer runs. You should not edit your registry unless it is absolutely necessary. If there is an error in your registry, your computer will become nonfunctional. If this happens, you can restore the registry to the state it was in when you last successfully started your computer.

TASK	STEPS

To restore the registry	Click on "Start"
	Click on "Shut down"
	Click on "Start the computer in MS-DOS mode"
	Click on "Yes"
	Change to the Windows directory
	Type the following commands and press <Enter> at the end of each command line:
	
	      Attrib -h -r -s system.dat
	      Attrib -h -r -s system.da0
	      Copy system.da0 system.dat
	      Attrib -h -r -s user.dat
	      Attrib -h -r -s user.da0
	      Copy user.da0 user.dat
	
	NOTE: the above procedure will restore the registry to the state it was in when you last successfully started your computer

Hints and Tips

TASK	STEPS
To simulate the Windows 3.1 program manager	Place the file PROGRAM.EXE in the startup folder

TASK	STEPS
To simulate the Windows 3.1 file manager instead of Explorer	Place the file WINFILE.EXE in the startup folder

TASK	STEPS
To close all windows	Hold down the <Shift> key and click on X

TASK	STEPS
To set the computer's clock	Double-click on the Taskbar

TASK	STEPS
To change all windows' color scheme	Right-click on the desktop, then properties, then Appearance.

TASK	STEPS
To change the screen saver	Right-click on the desktop, then properties, then screen saver

TASK	STEPS
To have a program start immediately when Windows starts	Drag the program icon into the startup folder

TASK	STEPS
To change the desktop background	Right-click on the desktop, then properties, then background

TASK	STEPS
To see a menu of available commands	Right-click anywhere

TASK	STEPS
To draw a selection box around a group of files	Click at the corner of the group, then drag to form the box

TASK	STEPS
To solve hardware problems	Use the Hardware Conflict Troubleshooter in Help

TASK	STEPS
To show or hide MS-DOS three-letter extensions	Set it in the Windows Explorer

TASK	STEPS
To boot in safe mode	Hold the <Shift> key down during bootup

Dual Boot Setup for Windows 95/98

You can have both Windows 95/98 and Windows 3.1 boot up from the same hard drive. You can also choose the operating system, and doublespace and drivespace drives you want to use when you first boot up. You can run your old programs using the old DOS/WIN system. Only use this method until you get used to Windows 95/98. Here is a step-by-step procedure on how to install a dual boot menu for Windows 95/98 and DOS 6/Windows 3.1:

1) Copy (duplicate) the whole Windows directory and all its subdirectories to another directory called WIN31. Also copy the DOS directory to another directory called DOS6.
2) Edit the INI files in the WIN31 directory and change all references from WINDOWS to WIN31. The "find and place" command in most word processors makes this easier. Make sure you save the changes as text files.

3) Boot up your PC and install Windows 95/98 SETUP through the Windows program manager "FILE-RUN". As you install Windows 95/98, it will rename your CONFIG.SYS and AUTOEXEC.BAT files to CONFIG.DOS and AUTOEXEC.DOS. (When you dual boot it renames them back to CONFIG.SYS and AUTOEXEC.BAT. Then it backs up the Windows 95/98 versions as W40).

4) After Windows 95/98 finishes installing and you are on the desktop, use Notepad with "select files *.*" to edit the CONFIG.DOS and AUTOEXEC.DOS files and change all WINDOWS references to WIN31 and all DOS references to DOS6.

5) In order for the dual boot menu to work, the following lines must appear in the [options] section of your MSDOS.SYS file. (If you use doublespace or drivespace disk compression you must change both MSDOS.SYS files: one on the boot drive and one in the compressed drive).

> [options]
>
> BootGUI=1
> Network=0
> BootMulti=1
> BootMenu=1
> Boot Menu default=7 (Original DOS default. Use 1 for Windows 95 default)
> BootMenudelay=5 (Number of seconds to select something else)
>
> Do the following to add the above steps:
>
> a) Open "My Computer"
> b) Use the menu bar to VIEW-OPTIONS-VIEW
> c) Show ALL files
> d) Unclick "Hide MS-DOS file extensions"
> e) Double-click on the MSDOS.SYS icon
> f) Edit the above lines if they do not exist

6) Since the MSDOS.SYS file is "read-only-hidden," you'll need to change the file attributes and then use the Notepad editor to insert the new line(s). Then you'll have to change the attributes back to their original states when you are done to protect the file.

> Do the following to change the attributes:
>
> a) Right-click on the MSDOS.SYS icon
> b) Select properties
> c) Unmark the attribute blocks
> d) Exit and make your text changes
> e) Save the file
> f) Put the attribute checks back where they were

7) Now exit Windows 95/98 and reboot. You should now see the boot menu and be able to select which mode you want to boot into.

Remember that any programs you install after this will appear only in the system you were running at install time. For example, if you install MS WORD under Windows 95, you will have to install it again under the old DOS/Windows if you want to run it both ways. You can install the program to the same directory both times and you'll simply overwrite the files and avoid having two sets of the new program on your hard drive.

Questions

1) What is the difference between Windows 95/98/2000 and Windows 3.x?
2) Explain what each of the following are:

    a) Icon
    b) Taskbar
    c) Vertical Scroll Bar
    d) Horizontal Scroll Bar
    e) Title Bar
    f) Menu Bar
    g) Toolbar

3) When do you use the restore button?
4) What happens when you click the minimize button?
5) How do you move a window around the desktop?
6) How do you resize a window?
7) Why can certain applications run under Windows 95/98/2000, but not under DOS 6.22?
8) What are the minimum processor and memory required for installing Windows 95/98/2000?
9) What is the "network neighborhood" feature in Windows 95/98/2000 used for?
10) What is the use of the Recycle Bin feature?
11) What is the use of My Briefcase feature?
12) List the steps to do the following:
    a) Copy random files from drive C to drive A
    b) Find the version of Windows 95/98/2000
    c) Use the volume control in the Taskbar
    d) Scandisk the hard drive
    e) Remove "Auto-Run" when a CD is inserted
    f) Rename the "My Briefcase" icon
    g) Free up space on the hard drive
    h) Format a disk in drive A
    i) Back up the entire system
    j) Hide the Taskbar
    k) Minimize all windows
    l) Add a new folder
    m) Set the computer's clock
    n) Simulate the Windows 3.1 file manager
    o) Recover files that have been deleted
    p) Change the wallpaper
    q) Select five files in consecutive order

13) What key(s) do you press to display the Windows 95/98 startup menu?

14) What is meant when Windows 95/98 is running in a safe mode?

15) What key(s) do you press to start Windows 95/98 in safe mode?

16) What is the registry?

17) What are the steps to restore the registry?

18) List all the steps needed to create a dual bootup.

5

Computer Service and Support

In this chapter, you will learn how to define, practice, and integrate the basic troubleshooting skills needed to become a successful computer technician. A computer technician should be able to service and support clients by configuring, installing, and troubleshooting hardware and software for any IBM-compatible PC.

Introduction to Service and Support

To be able to perform service and support on an IBM-compatible PC, a computer technician must be able to configure hardware, such as computer add-on boards, hard disks, and disk controllers. A computer technician must also be able to install various computer hardware components and troubleshoot or investigate whether a problem is related to software or hardware by diagnosing the problem; then once it is found, to fix the problem, and document the problem. Finally, the computer technician must be able to upgrade hardware and software as a means of supporting the operation of a PC. Here are some of the tasks a computer technician should be familiar with:

- Use a troubleshooting model to avoiding problems with electrostatic discharge. Document all system problems using diagnostic and system information software.

- Install and troubleshoot cabling and computer boards carefully. Examine and eliminate the problems that occur with various techniques on how to select, configure, and install computer boards.

- Use diagnostic software to troubleshoot common computer problems. Make sure to create a disaster recovery plan and learn the utilities used for disaster recovery.

- Prevent, diagnose, and resolve the common computer printing problems

- Configure, install, and troubleshoot various computer storage devices, including hard disks and CD-ROM drives. Be able to partition a hard drive.

Troubleshooting is part art and part science. Many attempts have been made to reduce troubleshooting to a set of procedures or flowcharts. However, no one can create a procedure or flowchart to deal with every possible problem. The key to troubleshooting is to develop the ability to break down a problem into its parts and see the interrelationships among those parts. It is the combination of knowledge and experience that will help us develop an efficient on-the-spot strategy to tackle each unique problem. Before you start, you should look for the common causes of problems before you do anything else. Depending on the type of problem we are facing, you may try the following steps first:

1) Eliminate user error	There are three possibilities when someone tells us that a network problem exists. Somebody tried to use a procedure and got a result different from what he or she expected, or got no result at all: the user did not do the procedure correctly; the procedure is working fine, but the user does not realize it; a problem exists. The way to determine which of these applies is to do the procedure once more, carefully thinking through each step. If the result is the same, we then ask ourselves the question, "How do I know this is not working correctly?"
2) Check the inventory	You should ask the following questions: Are all the parts, such as cables, present? Are they the right parts? Are they connected properly to each other?
3) Backup the disks	If disks are involved, we should back up the data before proceeding.
4) Cold boot the system	Turn the power off and turn it back on again.
5) Remove elements	Simplify the system by removing unneeded elements. For example, remove TSR programs that are running in the background and remove the AUTOEXEC.BAT and CONFIG.SYS files to free memory, Remove expansion slot boards one at a time to expose conflicts.
6) Check the application	Suspect a rights issue if you can run one application successfully, but have trouble running another application.
7) Check the software version	Make sure that the most current version of the software is being used.

You may use the following four steps for systematic troubleshooting:

- Gather basic information—Determine what the symptoms are and who on the network is being affected by the problem. Determine usage at the time when the problem occurs. Is the PC broken or just saturated with activity? PC analyzers, such as WINCheckIt for Windows, may be useful in verifying usage. Check PC logbooks and record-keeping devices to determine what the normal or baseline performance is for the PC. The activity and configuration information records may help us answer the critical question, "What has changed since the last time it worked?"

- Develop a plan to isolate the problem—Come up with two or three hypotheses using the basic information you have gathered and the background knowledge you have about the PC to decide whether the problem came from a user error, application software, or equipment. You should prioritize your hypotheses to decide which one to work on first based on: The likelihood that a given solution is the right one; the cost of trying the solution.

  Imagine that we have three hypotheses about a problem. Based on our experience, we guess hypothesis A to have a 50% chance of being the problem, hypothesis B a 30% chance, and hypothesis C a 20 % chance. We would probably want to try them out in the order of A, B, then C. Imagine the preceding scenario with one additional factor. Assume that hypotheses A and C will be time-consuming to test, but hypothesis B is a one-minute quick fix. Now it makes sense to try them out in order of B, A, C.

- Execute the plan—You should break down each hypothesis into the smallest reasonable, testable concepts.

  Test out the hypothesis by changing only one thing at a time. Otherwise, we will never know which change was the solution. After making one change, test the system to find out if the problem has been affected. When dealing with PC communication problems, start working at the source device and check the testable concepts by moving toward the destination device. Working from the source device toward the destination is called forward chaining. Working from the destination device towards the source is called backward chaining.

  You should use only reliable test equipment, software, and procedures. If you are testing a system with a faulty piece of equipment or utility disk, you will be more confused about what was wrong to begin with.

- Document the solution—You should document the solution and take steps to avoid or prepare for recurrence. You should record the nature of the problem and the solution in the PC logbook. This record will provide a quick fix if a similar problem occurs later on. The best predictor of future performance in a computer is its past performance. You should do what can be done to prevent or prepare for a recurrence of the problem. If a certain component burns out regularly, you may want to try installing a cooling fan. If prevention is impossible, you should replace the unit, or have a lot of spare parts handy.

Electrostatic Discharge Maintenance

Electrostatic charges are generated whenever two objects are joined together and then separated. Upon separation, the atoms of one object will take electrons and become negatively charged, while the atoms of the other object will give up electrons and become positively charged. Normal movements such as lifting a foot or moving a chair can generate charges in the range of 1,000 volts, which we do not notice. To be felt, the charge of an electrostatic discharge or ESD must equal about 3,000 volts.

As electronic components such as microchips have become smaller and denser, they have become susceptible to damage from ESD events of very small voltages. Computer components may be degraded or destroyed by discharge as low as 20 to 30 volts. ESD may cause failures in computer equipment during all stages of production, handling, shipping, and field maintenance. Direct physical contact is not required for static charges to build.

About 90% of the time, ESD events cause the component to degrade but not fail testing procedures, resulting in failure at a later date. Since components do not fail immediately, technicians often underestimate the cost of not using ESD prevention measures.

Practice ESD prevention and encourage your customers and suppliers to do so. There are many benefits of a good ESD control program as shown below:

a) Less need for spare hardware inventory

b)   Less downtime
c)   Fewer difficult-to-trace intermittent problems
d)   Fewer unnecessary service visits
e)   Fewer disgruntled customers

Follow the static prevention rules listed below to protect our equipment from ESDs:

1) Use a wrist strap and mat—Use the proper wrist strap and mat to ground yourself and your equipment before working on any device containing a printed circuit. But you should never use a wrist strap when working with monitors. They may carry a large voltage that can reach us through the strap. Test daily to make sure grounds have not become loose or intermittent.

2) Do not touch the leads of ICs—Never touch component or integrated chips by their electrical leads.

3) No conductive surface allowed—Never place components on any conductive surface.

4) Keep nonconductors away—Keep nonconductors, such as plastic, styrofoam, synthetic clothing, and polyester ties, away from open computers and components. Nonconductors are a source of static charges.

5) Keep people away—Do not let anyone touch you while you are working on boards that contain ICs. They may cause static charge.

6) Keep humidity @ 70-90%—Keep the humidity of any area with open computers at 70 to 90 percent. Static problems are much more likely to occur in low humidity.

7) Use static shielding bags—Always transport and store boards and ICs in static shielding bags. No pinholes are allowed in the bags since they will defeat the purpose. Static shielding bags and anti-static bags are not the same. Proper static shielding bags often have a gray-silver tone tint. Anti-static bags, usually tinted pink or blue, do not shield their contents from external static fields and should not be used.

Problem Prevention

Keeping necessary system documentation may save you many hours in troubleshooting problems. You may divide the system documentation into three categories:

a)   The PC system
b)   The history of the PC
c)   Resources used with the PC

Document the following to help you solve physical network problems:

1)   The LAN (Local Area Network) map—A detailed graphic record of the LAN that identifies the locations of all users, user groups, servers, printers, repeaters, bridges, gateways, and wiring centers.

2)   The LAN inventory—A record of the hardware, software, and peripherals used in the LAN.

You may purchase software that maintains records of the hardware, software, and peripherals on the system.

3) Cabling documentation—A record of the actual cabling, which wire goes where, in the LAN. Accurate documentation is critical in solving cable problems.

4) Workstation documentation—A record of each workstation configuration and its role in the network. The configuration and role of individual workstations can be critical to a network's health and should therefore be documented.

5) Chronological change log—A record of changes in the LAN configuration, including software upgrades and hardware that has been added or removed.

Document the history of the LAN as shown below to help you solve a problem that may occur more than once.

1) The business environment—Many network administrators concentrate too much on the technology and forget that the real system is the whole business environment. This document would answer questions such as the following:

   a) How is this LAN being used
   b) What is the relationship between the LAN and the company

These are important things to know in dealing with the human aspects of troubleshooting.

2) User information—This document would list user's names and the following information:

   a) Where the users are located
   b) What the users typically do on the system
   c) What training the users received

3) Baseline information and user patterns usage—This document indicates how the network functions under normal circumstances, including statistics on network traffic, CPU usage, bandwidth utilization, and errors and other information.

4) Log of past problems and trouble reports—This document would include details about what has happened over the past two years, including:

   a) Problem descriptions and resolutions
   b) Downtime incurred
   c) Performance analysis of devices such as printers and routers

Document the resources used with the LAN to help you track down resources or people who might know the answer to your problem:

1) Technology documentation—This document would include the protocols, routing, LAN architecture, and other technical materials specific to your network.

2) Technical support—This document should include:

   a) Key contacts in the vendor's service departments

b)   Current technical bulletin board numbers

The purpose of diagnostic software is to quickly provide information about your system that can assist you in isolating and solving problems. When acquiring a diagnostic utility, we should look for a package that:

a)   Works with many aspects of our system
b)   Is known to work with our hardware
c)   Has a good user interface
d)   Meets our reporting requirements
e)   Has adequate and timely support available

Use diagnostic software such as CheckIt PRO and WINCheckIt (the Windows 3.1 version), the system information tools for PCs, to accomplish the following tasks:

1)   Get system information—Get information on system hardware and operating system. The report becomes part of the LAN network record-keeping system..

2)   Take system inventory—Take inventory of the internal components of computer, such as the date of BIOS, type of CPU, and the amount of RAM.

3)   Edit and view CMOS—You may protect your CMOS against battery loss and restore CMOS to other machines with the same configuration. This can be useful in a large networked organization where many nodes may have identical configurations.

3)   Check IRQ, I/O, & ROM—Provide information on IRQs, I/O addresses, and memory addresses. This is to help us avoid incompatibilities when we add a network board or some other peripheral device to our system.

5)   Test hardware components—Test and diagnose the following components: system board, hard disk, floppy disk drive, serial port, parallel port, and memory.

6)   Check drivers and TSR—Collect system information, including device drivers and TSRs. This helps us to determine conflicts between software components.

7)   Benchmark system—Take benchmark readings on system performance. This allows us to compare our system to itself or to other systems. Preventive maintenance could include generating performance data at regular intervals and comparing it to earlier data.

In addition, WINCheckIt allows you to do the following:

1)   Enhance performance—It provides many system utilities to enhance workstation performance.

2)   A clean-up utility—It removes unneeded files.

3)   An uninstall utility—It identifies and deletes files for any Windows 3.1 program.

4)   A memory tune-up utility—It consolidates Windows 3.1 memory fragments and increases the largest memory block available.

5) Use of setup advisor—(Hardware compatibility) Use the setup advisor to report on the compatibility between existing hardware and a list of over 200 expansion boards and peripherals we may want to install in the system.

6) Use of software shopper—(Software compatibility) Use the software shopper to report on the compatibility between approximately 1,500 software files.

POST Audio and Error Codes

Let's start with basic troubleshooting. As you know, every time you turn your computer on, it automatically goes through a series of internal tests. These tests are called the Power-On-Self-Test (POST). The POST is contained in the computer's ROM. The sequence taken by the POST is as follows:

a) Basic System Test—Checks the operation of the CPU, system bus, and the memory segment containing the POST ROM.
b) Extended System Test—Checks the system timer and ROM BASIC interpreter.
c) Display Test—Checks the hardware that operates the video signals. It will test the video RAM for a default display adapter and the video signals that drive the display.
d) Memory Test—Checks the computer's memory using the writing and reading back of patterns.
e) Keyboard Test—Checks the keyboard interface and looks for stuck keys.
f) Cassette Test—Checks the cassette recorder interface of the PC.
g) Disk Drive Test—Checks to see what disk drives are installed. If there is a disk installed, then the disk interface card is tested.
h) Adapter Card Test—For PS/2 systems this test checks and configures an installed adapter card.

Generally, when the POST encounters an error on an IBM PC, it will indicate the type of error by a POST error code (number) or an audio sound. The POST error codes are numbers whose values indicate the type of problem you have encountered. Using the system speaker, The POST can indicate the type of problem by a series of short beeps.

If an error code of 161 is posted during the POST, it means that the battery on the motherboard which keeps the information in your CMOS (SETUP) program has died and needs to be replaced. When a computer is operated for the first time after it has been manufactured, the system has to be configured before it can be used. If the system battery fails and has to be replaced, the system must be reconfigured.

On some computers, the configuration is done by software on a diagnostic disk that comes with the computer system. You would follow the instructions to run the software and enter the information as asked for, such as date, time, type and number of floppy drives, type and number of hard drives, type of video adapter and monitor, and type and size of your memory.

On older computers, jumpers and switches had to be used to get the information above. All of this information is stored in a CMOS (BIOS) chip on the motherboard. This is the chip that is powered by the 12-volt battery. Table 5.1 shows the IBM BIOS audio and error codes and Table 5.2 shows the AMI BIOS and audio codes.

Audio and Numeric Error Codes	Probable Cause/Description
One short beep	Everything is normal
No beep, or short series of beeps	Power supply
One long beep and one short beep	Motherboard
One long beep and two short beeps	Video adapter card, monitor, or cable
One short beep, the system does not boot, and the drive light stays on	A disk drive, its adapter, or its interconnecting cable
161	Dead battery on the motherboard, SETUP (CMOS has been corrupted and must be reset)
131	For IBM PC, the cassette
601, 1780, or 1781	Disk drive
020 to 030	Power supply
100 to 199	Motherboard
200 to 299	RAM memory
300 to 399	Keyboard
400 to 499	Monochrome display, or on PS/2 systems the board parallel port
500 to 599	Color graphics adapter
600 to 699	Floppy drive or floppy drive adapter
700 to 799	Math coprocessor
900 to 999	Parallel printer port
1000 to 1099	Alternate parallel printer
1100 to 1199	A serial port
1200 to 1299	Alternate serial port
1300 to 1399	Game adapter
1400 to 1499	Dot-matrix printer
1700 to 1799	Fixed drive (Hard drive)
1800 to 1899	I/O expansion unit
2400 to 2499	EGA card, or on PS/2 systems, VGA card
2900 to 2999	Color or graphic printer
3000 to 3099	Primary network card
3100 to 3199	Secondary network card
3600 to 3699	General-purpose interface bus (GPIB)
3800 to 3899	Data acquisition adapter
7100 to 7199	Voice communication adapter
8500 to 8599	IBM expanded memory adapter
8600 to 8699	PS/2 point device (Mouse)
10400 to 10499	PS/2 Fixed disk (Hard drive)

Table 5.1 IBM BIOS major audio and error codes and their probable causes

Audio and Error Message Codes	Probable Cause/Description
One beep (DRAM refresh failure)	The memory refreshing circuitry on the motherboard is faulty
Two beeps (Parity error)	Parity error in the first 64 KB of memory
Three beeps (Base 64 KB or CMOS failure)	First 64 KB of RAM or CMOS memory failure
Four beeps (System timer failure)	Timer on the motherboard is not functioning
Five beeps (Processor error)	The CPU generated an error
Six beeps (8042 error or Gate A20 failure)	Keyboard controller (8042) may be bad. The BIOS cannot switch to protected mode
Seven beeps (CPU exception, interrupt error)	The CPU generated an exception, interrupt
Eight beeps (Display memory R/W error)	The video card is either missing or its RAM is bad
Nine beeps (ROM BIOS checksum error)	ROM checksum not equal to the value encoded in BIOS
Ten beeps (CMOS shutdown R/W error)	The shutdown register for CMOS RAM failed
Eleven beeps (External cache bad)	The external cache is faulty

Table 5.2 AMI BIOS major audio and error codes and their probable causes

Advanced Troubleshooting

With the new and more advanced software applications and high-speed IBM PC compatibles, a technician needs more troubleshooting techniques than the POST audio and error codes. The following are several problems that a PC technician might encounter and the possible causes with a course of action suggested to solve the problem. Note that the probable causes are prioritized.

Floppy and Hard Drive Problems

Problem 1:   Floppy drive stays on

Probable Cause: The floppy drive cable is not connected correctly

Course of Action: Reconnect the floppy drive cable and make sure that Pin 1 of the cable is connected to Pin 1 of the floppy drive

Problem 2:   Error reading drive A

Probable Cause: a) Bad floppy disk
b) Floppy disk is not formatted

Course of Action: a) Install a new floppy drive
b) Format the floppy disk

Problem 3:   C drive failure

Probable Cause: a) The SETUP (CMOS) program does not have the correct information about the hard drive.
b) The hard drive cable is not connected correctly.

Course of Action:	a) Boot the system from drive A and input the correct information in the SETUP program. b) Make sure that the cable is connected correctly
Problem 4:	If you cannot boot your system after you have installed a second hard drive.
Probable Cause:	a) The Master/Slave jumpers are not set correctly b) The hard drives are not compatible. They are from different manufacturers
Course of Action:	a) Set the Master/Slave jumpers correctly b) Call the drive manufacturer for compatibility with other drives. Run the SETUP program and select the correct drive types
Problem 5:	If a disk has been formatted on an IBM PS/2, it will not operate with compatible systems
Probable Cause:	The IBM PS/2 uses a different format than other computers
Course of Action:	Format a disk in an AT-type computer, insert the disk into the IBM PS/2, and copy the files you wish
Problem 6:	If the system does not boot from a hard disk but can be booted from a floppy disk
Probable Cause:	a) The connector between the hard drive and the system board is unplugged b) Damaged hard disk or disk controller c) Hard disk directory or the FAT table is corrupted
Course of Action:	a) If you tried to run the FDISK utility described in the hard disk section, you will receive an "Invalid Drive Specification." Check to see if the cable is running from the disk to the disk controller board. Make sure both ends are securely plugged in. Check the drive type in the standard CMOS (SETUP) b) Try to format the hard drive; if you are unable to do so, then the hard disk is defective. Try to replace the disk controller or contact technical support c) FDISK the hard drive and reformat it. Do not forget to copy all the data that was backed up onto hard drive. Backing up the hard drive is extremely important. All hard disks are capable of breaking down at any time
Problem 7:	If the system does not boot from a hard disk, but it can be booted from a floppy disk and you can read the applications from the hard drive
Probable Cause:	The hard disk boot program has been destroyed
Course of Action:	Anything could have caused this problem. Back up all the data on floppy

disks and reformat the hard drive. Reinstall applications and data using backup disks

Problem 8: If you receive the message "Sector Not Found" or other messages not allowing you to retrieve data

Probable Cause: Again, anything could have caused this problem

Course of Action: Use a file-by-file backup instead of an image backup in order to back up the hard disk. Make sure to backup all salvageable data. Then FDSIK and reformat your hard drive. Use SCANDISK to fix the disk surface

Problem 9: If you receive the message "ROM BASIC is not present"

Probable Cause: You did not activate the primary DOS partition

Course of Action: Activate the primary DOS partition

Monitor Problems

Problem 1: If you see the ASCII codes on the screen (Greek-looking letters)

Probable Cause: a) You have a memory problem
b) The display card is not set correctly
c) Computer virus

Course of Action: a) Reboot your system and reinstall the memory modules in the correct sockets
b) Check the jumper and switch settings on the display card
c) Format the hard drive

Problem 2: If the screen goes blank periodically

Probable Cause: The screen saver is enabled

Course of Action: Disable the screen saver

Problem 3: If you get no color on the screen

Probable Cause: a) Bad monitor
b) The SETUP program is not set correctly

Course of Action: a) See if you can connect the monitor to another PC; if there is still no color, replace it
b) Call for technical assistance

Problem 4: Screen is blank

Probable Cause: No power to the monitor, or the monitor is not connected to the computer, or the network card I/O address has a conflict

Course of Action:	Check the power to the system and the monitor. Also make sure that the monitor is connected to the display card. Change the I/O address on the network card if applicable

Keyboard Problems

Problem 1:	Keyboard failure
Probable Cause:	Keyboard is not connected correctly
Course of Action:	Reconnect the keyboard and check it again; if no improvement, replace it
Problem 2:	Certain keys do not function
Probable Cause:	The keys are jammed or defective
Course of Action:	Replace the keyboard
Problem 3:	The keyboard is locked and none of the keys function
Probable Cause:	Keyboard is locked
Course of Action:	Unlock the keyboard

BIOS Chip (CMOS SETUP Program) Problems

Problem 1:	If you receive the message "Missing operating system"
Probable Cause:	CMOS setup has been altered
Course of Action:	Run SETUP and select the correct drive type
Problem 2:	If you receive a message "Invalid configuration" or "CMOS failure"
Probable Cause:	The CMOS (SETUP) program does not have the correct information
Course of Action:	Review the system's equipment, check the configuration program and make sure to replace any incorrect information

PC Miscellaneous Problems

Problem 1:	If there is no power to the system at all, the power light does not illuminate, the fan inside the power supply does not turn on, and the indicator lights on the keyboard do not turn on
Probable Cause:	a) Defective power cable b) Defective power supply c) Bad wall outlet or the circuit breaker or fuse is blown
Course of Action:	a) Inspect the power cable and make sure that the power cable is securely

connected or try a new power cable; if it works, replace the old one
b) Check to see if the power cable and the wall socket outlet are OK first; if OK then replace the power supply or call for technical assistance
c) Plug the system in a different socket and test it. Repair the outlet and reset the breaker or replace the fuse

Problem 2: If the system is not operational, the keyboard lights stay on, the power indicator light is lit, and the hard drive is spinning

Probable Cause:
a) An expansion card is not connected correctly in an expansion slot on the motherboard
b) Defective floppy or tape drive
c) Defective expansion card

Course of Action:
a) Turn the system off. Check to see if all the cards are seated in their expansion slots. Press firmly on both ends of the expansion card to securely seat it in the expansion slot
b) Turn the system off, disconnect all the cables from the floppy drive, and turn your system back on. Check to see if the keyboard operates. Repeat until the bad drive is located
c) Turn off the system and replace the expansion card. Make sure it is secured in the expansion slot

Problem 3: If after you installed an expansion card, the system no longer operates

Probable Cause: IRQ conflict

Course of Action: All or part of the system may be inoperable. The new card may work, but a mouse or a COM port may not. Change the interrupt or RAM address on the new card. See the documentation for the card in order to change the settings

Laser Printer Problems

Problem 1: Printer stays in warm-up mode. A continuous "Warming Up" status code displays

Probable Cause: Faulty communication interface or a control panel problem

Course of Action: Turn the printer off, disconnect its communication cable, and restore power. If the printer finally becomes ready without its communication cable, check the cable itself and its connection at the computer. If the printer still fails to become ready, unplug the printer and check that the control cables or interconnecting wiring is attached properly.

Problem 2: The message "Paper Out"

Probable Cause: Either the paper is exhausted or the paper tray has been removed

Course of Action:	Remove the paper tray. Be sure that there is paper in the tray and that any ID tabs are intact. You can check the paper ID microswitches by removing the paper tray and actuating the paper-sensing arm by hand (so the printer "thinks" that paper is available)
Problem 3:	The message "Printer Open"
Probable Cause:	Printer door is open
Course of Action:	a) Make sure that all covers are shut securely (try opening and closing each cover) b) Unplug the printer and use your multimeter to measure continuity across any questionable interlock switches. It may be necessary to remove at least one wire from the switch to prevent false readings. c) If the switch itself works correctly, check the signals feeding the switch
Problem 4:	The message "No EP Cartridge"
Probable Cause:	The printer does not see the EP cartridge
Course of Action:	a) Check the installation of your current EP cartridge. Make sure that the cartridge is in place and seated properly b) Check to be sure that at least one tab is actuating the sensor switch c) Unplug the printer and use your multimeter to measure continuity across each sensitivity switch
Problem 5:	The message "Toner Low" appears constantly, or the error never appears
Probable Cause:	The toner sensor
Course of Action:	a) Shake the toner to redistribute the toner supply (or insert a fresh EP cartridge). b) Repair or replace the high-voltage power supply. Use extreme caution when attempting a high-voltage repair! Allow plenty of time for the supply to discharge before disassembling the printer. If you must replace components, make sure to use parts with the proper voltage ratings
Problem 6:	The message "Paper Jam" appears
Probable Cause:	Several different combinations of conditions
Course of Action:	a) Check the paper tray. If a jam condition is shown but there is no paper, the paper sensing arm is not functioning properly. It may be broken, bent, or jammed. b) If there is ample paper available, take a moment to be sure that the paper is the right size, texture, and weight (bond) for your printer c) Unplug the printer, open its covers, and carefully note the paper's position. Look closely at the leading and lagging paper edges. The problem can often be isolated based upon the paper's jam position.

Problem 7:   The general scanner "error" message

Probable Cause: Main scanner motor

Course of Action:
a) Unplug the printer, open its housing, and carefully inspect connectors and interconnecting wiring between the motor and its driver circuits. Reseat any connectors or wiring that appear to be loose.
b) Troubleshoot the excitation voltage and switching circuitry back into the main logic board. Remember that you might have to defeat cover interlock to enable the printer's low-voltage power supply.

Problem 8:   The "Service" error indicating a fusing malfunction

Probable Cause: The fusing assembly

Course of Action:
a) Examine the installation of your fusing assembly. Check to see that all connectors and wiring are tight and seated properly. An ac power supply is often equipped with a fuse or circuit breaker that protect the printer. If this fuse is open, replace the fuse, or reset the circuit breaker, then retest the printer.
b) Unplug the printer and check the temperature sensor thermistor by measuring its resistance with a multimeter. At room temperature the thermistor should read about 1 kW (depending on the particular thermistor).

Problem 9:   Image formation problems

Probable Cause: Anything

Course of Action:
Although print quality is always a subjective decision, certain physical characteristics in ES printing signal trouble in image formation. It is virtually impossible to define every possible image problem, but from the basic symptoms you can tell where to look for trouble.

Problem 10:   Pages are completely blacked out or appear blotched with undefined borders.

Probable Cause: Primary corona wire

Course of Action:
a) Unplug the printer, remove the EP cartridge, and examine its primary corona wire. A failure in the primary corona prevents charge development on the drum.
b) If the blacked-out page shows print with sharp, clearly defined borders, the writing mechanism may be running out of control. In this case the primary corona is working fine. Use the oscilloscope to measure the data signals reaching the writing mechanism during a print cycle. You should observe a semirandom square wave representing the 1's and 0's composing the image. If you find only one logic state, troubleshoot your main logic and the driving circuits handling the data. If the data entering the writing mechanism appears

normal, replace the writing mechanism.

Problem 11:   Print is very faint

Probable Cause: Toner in the EP cartridge

Course of Action:
a) Unplug the printer, remove the EP cartridge, and try redistributing toner in the cartridge by rocking the cartridge back and forth gently.
b) Check the transfer corona. The transfer corona applies a charge to paper that pulls toner off the drum. A weak transfer corona might not apply enough charge to attract all of the toner in a drum image. The result can be a very faint image.
c) Check the drum ground contacts to be sure that they are secure. Dirty or damaged ground contacts will not readily allow exposed drum areas to discharge. Very little toner will be attracted, and only faint images will result.

Problem 12:   Print appears speckled

Probable Cause: Fault in the primary corona grid.

Course of Action:
The grid is a fine wire mesh between the primary corona and drum. A constant voltage applied across the grid serves to regulate the charge applied to the drum to establish a more consistent charge distribution. Grid failure will allow much higher charge levels to be applied unevenly, resulting in dark splotches in the print. Since the primary grid assembly is part of the EP cartridge, replace the EP cartridge and retest the printer. If speckled print persists, repair the high-voltage power supply assembly.

Problem 13:   One or more vertical white streaks appear in the print

Probable Cause: Toner may be distributed unevenly.

Course of Action:
a) Check the toner level. Unplug the printer, remove the EP cartridge, and redistribute the toner.
b) Examine the transfer corona for areas of blockage or extreme contamination. Clean the transfer corona very carefully with a cotton swab. If your printer comes with a corona cleaning tool, use that instead. Be careful when cleaning, making sure to avoid the monofilament wrap around the transfer corona assembly. If the line breaks, the corona will have to be rewired, or the entire transfer corona assembly will have to be replaced.
c) A laser writing mechanism might be afflicted with a dirty scanning mirror. Like lenses, a scanning mirror can be dusted off with ultra-clean compressed air or cleaned with quality lens cleaner and photographic wipes. Be careful not to shift the mirror on its mountings.

Problem 14:   Right-hand text is missing or distorted.

Probable Cause: Low toner in the EP cartridge.

Course of Action:	a) Unplug the printer, remove the EP cartridge, and redistribute the toner.
	b) Examine the mountings that support your writing mechanism. If these are bad, try to replace the writing mechanism.
	c) If you are using a laser writing mechanism, pay special attention to the installation and alignment of the scanner mirror assembly. An end portion of the image might be distorted or missing. Reseat or replace any incorrectly positioned mirror.

Problem 15: Image registration is consistently faulty.

Probable Cause: Poor paper quality, mechanical wear, and paper path obstructions.

Course of Action:

a) Inspect the paper and paper tray assembly. Correct any damage or restrictions that you find or replace the tray.

b) If the registration is still incorrect, mechanical wear in the paper feed assembly is typically the problem. Check the pickup roller assembly first. Look for signs of excessive roller wear. The recommended procedure is simply to replace the pickup assembly; however, you might be able to adjust the pickup roller or clutch tension to somewhat improve printer performance.

c) If the set of rollers does not grab the waiting page and pull it through evenly at the proper time, misregistration can occur. As you initiate a printer self-test, watch the action of the registration rollers. They should engage immediately after the pickup roller stops turning. Otherwise, the recommended procedure is to replace the registration assembly, but you might be able to adjust torsion spring tensions to somewhat improve printer performance.

d) Dirty or damaged gears can jam or slip causing erratic paper movement and faulting registration. Clean the drive train gears with a clean, soft cloth. Use a cotton swab to clean gear teeth and tight spaces. Remove any objects or debris that might block the drive train, and replace any gears that are damaged.

Problem 16: Horizontal black lines are spaced randomly through the print

Probable Cause: Laser writing mechanism

Course of Action:

a) If your printer uses a laser writing mechanism, a defective or improperly seated beam detector could send false scan timing signals to the main logic. Reseat or replace the beam detector and optical fiber.

b) A loose or misaligned scanning mirror can also cause this type of problem. Vibrations in the mirror might occasionally deflect the beam around the detector. Realign or replace the scanning assembly.

Problem 17: Print is slightly faint

Probable Cause: Anything

Course of Action: a) Check the contrast control dial. Turn the dial to a lower setting to increase contrast (or whatever darker setting there is for your particular printer).

b) If the contrast control has little or no effect, the high-voltage power supply is probably failing. Repair or replace the high-voltage power supply.

c) Check the paper supply. Unusual or specially coated paper may cause fused toner images to appear faint. If you are unsure about the paper currently in the printer, insert a good quality, standard-weight xerographic paper and test the printer again.

d) Check the toner level. Unplug the printer, remove the EP cartridge, and redistribute the toner.

e) Unplug the printer and examine the EP cartridge sensitivity switch settings.

Problem 18:   Print has a rough or suede appearance.

Probable Cause: Serious failure in the main logic system.

Course of Action: Use the oscilloscope to trace the print data signal from the writing mechanism into the logic circuitry. It is best to just replace the main logic board.

Problem 19:   Print is smeared or fused improperly.

Probable Cause: The fusing assembly.

Course of Action:

a) Run a number of continuous self-tests (the printer does not have to be disassembled). After 10 printouts, place the first and last printouts on a firm surface and rub both surfaces with your fingertips. No smearing should occur. If the fusing level varies between pages (one page smears while another does not), clean the thermistor temperature sensor and repeat this test.

b) A cleaning pad rubs against the fusing roller to wipe away any accumulations of toner particles or dust.

c) Inspect the drive gears that show signs of damage or excessive wear. Slipping gears could allow the EP drum and paper to move at different speeds causing portions of the image to appear smudged (bolder or darker portions). Replace any defective gears. If you do not find any defective drive train components, replace the EP cartridge.

d) Finally, a foreign object in the paper path can rub against a toner powder image and smudge it before fusing. Check the paper path and remove any debris or paper fragments that might be interfering with the image.

Problem 20:   Printed images are distorted.

Probable Cause: Registration or scanning assembly.

Course of Action:

a) Image size distortion is indicated when characters appear too large or too small in the vertical direction. Large (or stretched) characters suggest that some portion of the pickup or registration assembly is failing or that some obstruction is in the paper path causing excessive drag on the paper. Small (or squashed) characters suggest a main motor problem—the motor may be moving the drum too fast.

b) Examine the pickup and registration assemblies for signs of unusual wear and replace any rollers or other mechanical parts that appear worn or damaged. Also inspect the EP cartridge. If the cartridge is very new or very old, try a replacement cartridge. If characters appear compressed, check the main motor and motor drive signals. Finally, look for debris or obstructions that might be interfering with drive train operation. Remove obstructions immediately.

c) Scanning distortion (typically found in laser printers) is indicated by wavy, irregularly shaped characters. This wavy distortion can also be seen in page margins. In most cases a marginal scanning motor causes minor variations in scanning speed (the motor speeds up and slows down erratically). Therefore, although scanning assembly may appear correct, replace the scanning assembly.

Problem 21:   Print shows regular or repetitive defects.

Probable Cause: EP cartridge or fusing roller assembly.

Course of Action: A dirty or damaged fusing roller may cause this type of problem. Unplug the printer, allow at least 10 minutes for the fusing assembly to cool, then gently clean the fusing rollers. If you find that the fusing rollers are physically damaged or you are unable to clean them effectively, replace the fusing roller assembly.

Inkjet and Dot Matrix Printer Problems

Problem 1:   Print quality is poor. Print appears smeared, faint, or smudged.

Probable Cause: Characteristics of paper used.

Course of Action: a) Inspect the ink supply and ink nozzles. A low ink supply or partially clogged nozzle(s) can cause the nozzle(s) to spit or sputter, resulting in a smudged appearance. Follow procedures to clean, prime, and retest the print head. Replenish your ink supply or replace the ink cartridge if necessary.

b) Check the print head spacing. If the print head is too close to the paper, ink impacting the page may spatter or run. A head that is too far away may produce print that appears to sag across a printed line. Most printers allow head spacing to be adjusted by several thousandths of an inch using a mechanical level assembly.

Problem 2:   Print contains one or more missing lines. This symptom also occurs during self-test.

Probable Cause: Ink supply

Course of Action: a) Check the ink supply. If the supply is marginally low, there may not be enough ink to supply every nozzle evenly. Nozzles that do not receive ink will not fire even if they are working properly. Replacing the cartridge is the best solution.

b) Missing lines can be caused by a fault in the print head cable. Intermittent or broken signal lines can disable any of the nozzles. Unplug the printer and use your multimeter to check continuity across each cable conductor. You may have to have to disconnect the print head cable at one end to prevent false continuity readings. Replace the print head cable that you find to be defective.

Problem 3: Print contains black lines. Symptom also occurs during self-test.

Probable Cause: Print head cable.

Course of Action: Replace the print head cable or connectors if they appear shorted.

Problem 4: Printer does not print under computer control. Operation appears correct in printer's self-test mode.

Probable Cause: Paper or communication cable interface.

Course of Action: a) Make sure that your printer is actually on-line with the host.
b) Paper might be exhausted.
c) The communication cable might have become loose or detached at one end.
d) Examine the DIP switch settings next. DIP switches are used to change optional settings such as serial communication format, character sets, or automatic line feed.

Problem 5: Print is intermittent or absent. All other functions appear correct.

Probable Cause: Low ink supply or trouble in ink flow.

Course of Action: a) Inspect the nozzles and ink flow carefully. If low ink supply is not able to provide adequate ink to all nozzles at times, you might see randomly occurring areas or missing dots.
b) Replaceable ink jet cartridges can usually be cleaned and primed easily. Use a clean swab dipped lightly in ethyl alcohol to wipe the face of each nozzle, then use the swab's wooden end to push gently on the ink bladder. This pressure forces fresh ink through each nozzle. When you see ink bead up in every nozzle, the print head is ready for use.
c) A faulty head cable can cause intermittent or missing print. Unplug the printer and use your multimeter to check continuity across each print head signal wire. You may have to disconnect the cable from at least one end to prevent false readings. Wiggle the cable to simulate any intermittent connections. Replace any defective print head cable.

Questions

1) What are the systematic troubleshooting steps?
2) What are the benefits of a good ESD control program?
3) What are the common causes of problems you should look for before you do anything else?
4) What are the static prevention rules
5) When is the POST executed?
6) Name all the internal tests that a POST executes and what each one is responsible for.
7) List all the possible causes and give a brief description of what action to take to solve the following problem scenarios:

   a) The POST posted an error code of 351
   b) The POST posted an error code of 161
   c) The POST indicated an audio sound of one beep
   d) The POST indicated an audio sound of one long beep and one short beep
   e) The POST posted an error code of 1100
   f) There is no color on the monitor
   g) You can boot the system from a floppy drive, but not from the hard disk
   h) The message "Sector not found" is displayed on the screen
   i) The message "ROM BASIC is not present" is displayed on the screen
   j) Greek letters are all over the screen
   k) The floppy drive light stays on
   l) The message "C drive failure" is displayed on the screen
   m) The message "Error reading drive A" is displayed on the screen
   n) You typed the command WIN at C prompt to load Windows and Windows did not execute
   o) You cannot hear the fan in the hard drive spinning when you turn the PC on

8) Describe the operation of the laser and inkjet printers.

6

Computer Networks

This chapter summarizes the concepts of computer networks. It will cover their services and protocols. This chapter will discuss the technology behind computer networks. It will cover topics such as network basics, network services, and the seven layers of the OSI (Open System Interconnection) model. In addition you will learn more about network transmission media, transmission devices, and network structures (network topologies).

Network Basics

A network is a group of devices that are connected together to allow them to communicate and to share files, printers, and other resources. Almost every company that has at least ten or more employees has some sort of workstations or clients connected to a server to form a network. A server on a network is a special computer that has resources used by clients and acts as the host for the other computers (workstations). The difference between a server and a workstation is that a server runs special software called a Network Operating System (NOS). The NOS is what makes a computer a server. The NOS provides you with two things. One of these things is resource sharing with security and managing users, such as creating, removing, or providing privileges to users. The NOS ties all the PCs and peripherals in the network together and coordinates the functions of all PCs and peripherals in a network. Finally, it provides security for and access to all the databases and peripherals in a network.

The NOS setup consists of two major multitasking operations—preemptive and non-preemptive. In preemptive multitasking, the NOS can take control of the processor without the task's cooperation, while in non-preemptive multitasking (cooperation), the task itself decides when to give up the processor. No other program can run until the non-preemptive program gives up control of the processor.

On the other hand, a workstation is a computer that connects to, requests data from, and exchanges information with a server. Often a workstation is referred to as a client. In fact, any peripheral that requests data from or exchanges data with a server is called a client. The client software performs several tasks:

1) Processes the forwarding of requests.

2) Intercepts incoming requests.

3) Determines if the incoming requests should be left alone to continue on the local PC or redirected out of the network to another server or peripheral.

4) Designates shared drives in other PCs.

A peripheral is any device, other than a computer, connected in a network or to any other computer. Every client, including the server, must connect to a network through a device called a Network Interface Card (NIC). The NIC is an interface card located inside a PC to make it interface with the rest of the network. Below is a list of tasks an NIC card performs:

1) Prepares the data from the PC to be placed on the network cable.

2) Sends the data to another PC.

3) Controls the flow of data between the PC and the cabling system.

4) Contains firmware programs that implement the logical link control and access media control functions (data link layer).

5) The NIC is responsible for making sure that data are transmitted serially. On the network cable data must travel in a single bit stream. This is called serial communication, which means that one bit follows another (like cars on a one-lane highway). Data can go one way at a time, either send or receive.

6) Translates the PC's digital signals to electrical and optical signals for the network cable (transceiver).

7) The network address on the NIC (regulated by the IEEE committee) is used for moving data from PC to PC. It also signals the PC requesting data for receiving. Finally the address is required for receiving the data from memory into the card memory (buffer). This address is called the MAC (Media Access Control) address.

8) The card enables you to select an Interrupt Request (IRQ) number (commonly IRQ 5 is free), and a base I/O port address. The I/O port is where information flows between the PC's hardware and its CPU. The port appears to the CPU as an address. In addition, you need to select a base memory address on the card to identify a location in RAM as a buffer area. Finally, you need to select what type of transceiver you will use (i.e. 10 Base2, 10 BaseT, or external).

The card may come in different architectures, such as ISA 8 bits, EISA 16 or 32 bits, Micro-Channel 16 or 32 bits, or PCI 32 or 64 bits. Data may be transmitted faster, depending on its architecture.

9) The card may come in different connectors, such as a BNC connector (round, thinnet), AUI connector (15 pins, like a joystick port, thicknet), or RJ-45 (like a telephone plug, 10 BaseT).

10) The speeds of the data coming in and going out of the card are determined by the card's DMA, buffering RAM, and the onboard microprocessor. DMA (Direct Memory Access) is an onboard chip responsible for moving data directly from the network card to the PC memory. The computers always share the card's memory and vice versa. When the card takes temporary control of the PC's bus in EISA and Micro-Channel architectures, it is called mastering.

Computer networks are generalized into three different computing categories: centralized computing, distributed computing, and collaborative computing. Centralized computing is generally handled by mainframes. A mainframe is a big computer that handles all the data storage and processing. It uses dumb terminals for input/output devices. Although data is not really shared, it is still called a network because of the connection of the dumb terminals to the mainframe. Centralized computing was the sole type of network before the evolution of personal computers in the 1980's.

With the evolution of personal computers, the distributed computing concept came into play. Distributed computing is a model that involves intelligent computers called clients. In a distributed computing model, a client has the power to process its own tasks and to exchange data and services with other computers. Collaborative computing (sometimes called cooperative computing), on the other hand, is an extension of the distributed computing model. In a collaborative computing environment, clients have internal processing power yet can still exchange data and services. The difference between collaborative and distributed computing is that collaborative computing allows multiple clients to perform tasks in order to maximize processing power.

Another way of categorizing networks is by their size. We can categorize a network according to its size in three different areas: Local Area Network (LAN), Metropolitan Area Network (MAN), and Wide Area Network (WAN). A LAN network is a network that usually covers a small area (i.e., a single building). In fact, any two PCs connected together will make up a LAN network. Typically in a LAN network, computers are connected to each other with a cable. A MAN network usually covers a metropolitan area and is a little bigger than a LAN, but smaller than a WAN. Different types of transmission media are required to connect PCs in a MAN network. Typically in a MAN network, computers are connected to each other with a transmission media such as cables, radio frequencies, or fiber optic cabling. The WAN network is the biggest of the three categories. A WAN network can cover a multicity or even a multination area. There are two types of WAN networks: enterprise networks and global networks. In an enterprise network, all of the clients are connected for a single organization or company, while in a global network, all of the clients are connected for multiple organizations that span the world. The Internet is an excellent example of a global network.

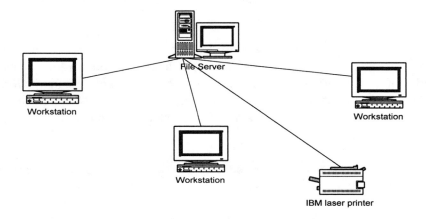

Figure 6.1 A typical LAN network

Network Services

For a network to work properly, it needs certain components such as network services, transmission media, and protocols. The reason we connect computers together in the first place is for them to share services. A computer on a network that provides services, such as a file server, is called a service provider. A computer that requests services is called a service requester. The service provider and all service requesters must be connected together in some sort of transmission media to enable them to share services. A transmission media is the physical cables or wireless technology that allows clients to communicate with one another. On the other hand, network protocols are the rules or guidelines of communications that allow computers and other peripherals to communicate with one another. We will discuss only network services in this section. Network transmission media and network protocol suites are discussed later in this chapter.

The five most common network services are file services, print services, message services, applications, and database services. File services are services that allow a user to transfer files, store and move files, back up files, and update synchronization of a file. In a network environment, you do not have to save your file on a disk to move it from one computer to another. Rather, you can save it on a network drive where any client (with security in place) can share and use that file. You can even exchange files with a server if you want to. File transfer services give you the ability to move or store big files on the network rather than on a workstation. A network can provide you with three types of storage categories. On-line storage gives you the ability to store data on a local hard drive, whereas off-line storage provides a means of storing data on a removable media such as a tape drive or an optical disk. Finally, near-line storage is when a machine such as a jukebox or a tape carrousel automatically retrieves and mounts the tape or a disk. File services also provide you the opportunity to back up your files and use file update synchronization to ensure that each user has the latest version of a specific file.

The second most common network service is print services. A network gives clients the opportunity to share printers, place printers where they are convenient with respect to all workstations, and allow clients to share network fax services. The network uses what is called a print queue to store print jobs before they are sent to a printer. Using a print queue, a network administrator can prioritize, move, or purge any print job from a print queue before that job is sent to the printer.

The third most common network service is message services. While file services transfer data in a file form, message services allow you to transfer data in any other form such as graphics, video, and audio. There are four types of message services: electronic mail, workgroup applications, object-oriented applications, and directory services. Now you can transfer messages on a network using electronic mail (e-mail). E-mail can transmit a text, audio, video, resources, and services on a network.

The fourth most common network service is application services as well as graphics. Workgroup applications, on the other hand, are used to produce more efficient ways to process a task among multiple users. You can handle a large number of tasks by using object-oriented applications message services. Object-oriented applications are programs made up of small applications called objects. Finally, a directory service is a directory or a database that contains all the information about the services. An application service on a network allows you to have a server dedicated to provide only application services. That is why a Windows NT server is an excellent application server. NetWare provides server applications by using NetWare Loadable Modules (NLMs) to support third-party applications.

The fifth most common network service is database services. Most database systems run under client/server base systems. Client/server base system implies that the server handles the intensive part of the task performance. With database services, you can distribute a huge database into portions to maximize network efficiency. NetWare 4.11 uses NDS (NetWare Directory Services) as its primary database management.

In 1977 the International Standards Organization (ISO) was formed to develop standards on a wide range of topics. The United States' representative of ISO is the American National Standards Institute (ANSI). ISO is most famous for its development of the Open System Interconnection (OSI) reference model. This model is designed to give developers a set of rules or guidelines to follow, and it organizes network communications into seven different layers: layer1- Physical, layer 2- Data Link, layer 3- Network, layer 4- Transport, layer 5- Session, layer 6- Presentation, and layer 7- Application. To help you remember the names and order (from layer 7 to layer 1) of the seven layers use the phrase "**A**ll **P**eople **S**eem **T**o **N**eed **D**ata **P**rocessing". Even though a description of each layer will be provided in the following sections, let's first take a look at a brief summary of how data are sent within the OSI model, and what responsibilities the protocols have in that model.

The functions of each layer in the OSI model communicate and work with the functions of the layers immediately above and below that layer. In addition, layers are separated from each other by a boundary called an interface. The purpose of each layer is to provide service to the next higher layer. Data are passed from one layer to another in the form of broken data called packets. If data were to be sent in large units, then the network would be tied up. Packets are therefore used to reduce traffic and errors in transmission. The packets consist of six components—source address, data, destination address, instructions for the network components on how to pass the data along, information that tells the receiving PC how to connect the packet to other packets and reassemble the complete data, and error-free checking.

You can also think of a packet as consisting of three parts—headers, data (contains CRC), and a trailer protocol.

Header	Data

The headers and CRC are attached or removed at each layer as they move within the OSI model. The headers are used to send an alert signal to indicate that a packet is being sent. In addition, a header consists of source address, destination address, and clock information. Data is the actual useful data, which can be in the range of 0.5 KB to 4 KB. The CRC is used to ensure the data integrity as it travels through the transmission media. The transport layer is where the original block of data gets divided into packets. In addition, sequencing information and address information are placed with the data as headers to guide the receiving PC in reassembling the data from the packet.

Again, network protocols are rules of behavior, which means that they are sets of procedures performing each task before sending data from one PC. Keep in mind that there are many protocols out there, and that some work at various OSI layers. Here are some facts about protocols and what they are responsible for:

1) The receiving PC uses protocols to break outgoing data into small sections (packets), add addressing information, and prepare the data for actual transmission.

2) The sending PC uses protocols to take incoming data packets off the cable, strip the addressing information, copy the data packets to a buffer for reassembly, and then pass the reassembled data to the application.

3) Several protocols may work together (protocol stack) in an OSI model:

    a) Application layer—The protocols initiate the request or accept the requests.

    b) Presentation layer—The protocols add, format, display, and encrypt information into the packet.

    c) Session layer—The protocols control the flow of information and send the packets on their way.

    d) Transport layer—The protocols add error-free handling information.

    e) Network layer—The protocols add sequences and address information.

    f) Data link layer—The protocols add error-check information (CRC) and prepare the data for the physical layer.

    g) Physical layer—The protocols packet and send the data as bit streams.

4) All protocols must be bound. For example, if TCP/IP is bound as the first protocol, then TCP/IP will be used to attempt to make a network connection. If this network connection fails, the PC will transparently attempt to make a connection using the next protocol in the binding order.

5) There are three basic types of protocols:

    a) Application protocols—These types of protocols work in the upper three layers of the OSI model. They can be application-to-application or data exchange protocols. Examples are SMTP or FTP.

    (E-mail) Simple mail transport protocol,

    b) Transport protocols—These types of protocols work at the transport layer. They provide for communication sessions between PCs and ensure that the data are able to move reliably. Examples are TCP, SPX, and NetBEUI (NetBIOS Extended User Interface, not routable).

    c) Network protocols—These types of protocols work at the lower three layers of the OSI model. They provide the link services and handle addressing, routing information, error checking, and retransmission requests. Examples are IP, IPX, Nwlink, and NetBEUI.

Transmission Media

Most analog and digital signals are transmitted through a bounded transmission media (cables). The capacity of a transmission media is referred to as the bandwidth. Bandwidth can be divided into channels to transmit more than one signal on a bounded transmission media. There are two types of transmission media bandwidth: baseband and broadband. Baseband transmission uses the entire media bandwidth to transmit digital signals on a single channel, with digital signaling over a single frequency. A digital signal uses the entire bandwidth of the cable (one channel). It is bidirectional, and uses a repeater to regenerate the signal. Most LANs use baseband to transmit their signals.

Broadband transmission, on the other hand, divides the media bandwidth into channels to support multiple, simultaneous signals over a single transmission medium. It is an analog signal, and it uses a wide range of frequencies. Broadband is also unidirectional. To solve this problem, it either uses a mid-split broadband, which divides the bandwidth into two channels with different frequencies, or uses two physical cables, one for sending, and the other for receiving. It uses amplifiers to regenerate the signal.

Multiplexing is a technique used to allow both baseband and broadband media to use multiple data channels. There are three methods of multiplexing: frequency-division, time-division, and statistical-time-division multiplexing. Frequency-division multiplexing (FDM) is a technique that uses different frequencies to add multiple data channels onto a broadband medium. Time-division multiplexing (TDM) is a technique that divides a channel into time slots. Each device communicates in its own allocated time slot. Statistical-time division multiplexing (StatTDM) is a technique that allocates time slots.

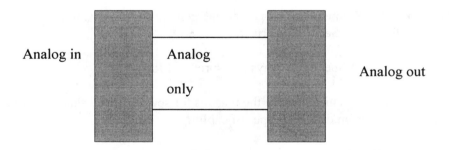

Figure 6.2 FDM

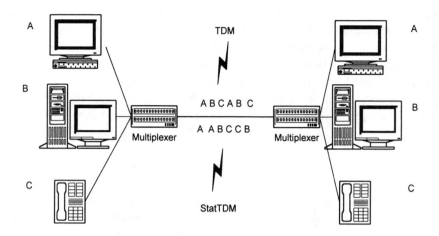

Figure 6.3 TDM and StatTDM

Signals must be transmitted through a transmission media. Transmission media come in two different classes: bounded and unbounded. Bounded media is a media that carries data in a physical pathway (i.e., cable). Unbounded media is a media that carries data through space (i.e., microwaves). Before we look at the four types of bounded media, UTP, STP, COAX, and fiber optic cables, we need to look at the considerations for cable:

1) Installation Logistics—How easy is it to install and work with the cable?

2) Shielding—The noisier the area, the more shielding the cable requires, thus increasing the cost.

3) Crosstalk—Remember that power line generators, motors, relays, etc. may crosstalk with the cable. Security is crucial.

4) Transmission Speed—Always measured in Mbps.

5) Attenuation—This is one of the reasons for specifications that recommend certain length limits on different types of cabling.

6) Cost—Of course this is also an important factor.

Unshielded Twisted Pair (UTP) is a cable that has two wires twisted together to reduce Electromagnetic Interference (EMI). It consists of two insulated strands of copper wire twisted around each other. The twisting cancels out electrical noise from adjacent pairs and from other sources, such as motors, relays, and transformers. UTP can be category 3, which means the cable has a four-twist-pair with three twists per foot. This makes it capable of transmitting up to 10 Mbps. If it is category 5, then it transmits up to 100 Mbps. UTP is the least expensive (because cables might already be installed). On the other hand, UTP is very sensitive to EMI, has a very low bandwidth, and is sensitive to crosstalk. UTP can be used in Token-Ring, Ethernet, or ARCnet networks.

Shielded Twisted Pair (STP) is a UTP cable with a braided shield placed around the twisted pair of wires to reduce EMI. Although EMI sensitivity in STP cable is low, the cable cost is a lot more expensive than that of a UTP. Even though the bandwidth is bigger than that of a UTP cable, STP cable is used only by IBM Token-Ring and Apple's LocalTalk networks. STP cable has a copper braid jacket, which is a higher quality, more protective jacket than UTP has. STP also uses a foil wrap between the wire pairs and internal twisting of the pairs. It is less susceptible to electrical interference, and thus a higher bandwidth is transmitted over a long distance. Both UTP and STP can travel up to 100 meters and they both need connection hardware, such as RJ45, which is similar to the RJ11, which is a telephone connector. Below are the five categories UTP and STP uses:

Category	Used For	Speed
1	Voice	< 4 Mbps
2	Voice	4 Mbps
3	Data	10 Mbps
4	Data	16 Mbps
5	Data	100 Mbps

A coaxial cable (COAX) is a cable that contains two conductors with a COmmon AXis. COAX is relatively inexpensive compared to a fiber optic cable. It is solid copper surrounded by insulation (a braided metal shielding). It can be dual shielded—where one layer is foil insulation and another layer is a braided metal shielding—or it can be quad shielded, where two layers are foil insulation and two layers are braided metal shielding. COAX is used for environments that are subject to higher interference. The shielding protects the transmitted data by absorbing stray electronic signals called noise. The insulating layers protect the core from electrical noise and crosstalk signals that overflow from an adjacent wire. The conducting core and the wire mesh must always be separated from each other (or have insulation in between them); otherwise, the cable will experience a short or noise.

COAX cables come in two types—thin (thinnet) 10Base2 or thick (thicknet) 10Base5. The thinnet cable can transmit data up to 180 meters without any attenuation. It uses RG-58 family cables with 50-ohm impedance. It can use RG-58 A/U cable, which is a stranded wire core, or use RG 58 /U cable, which is a solid copper wire. The thicknet wire is a standard Ethernet cable, which has a thicker copper core. A thicknet cable can transmit data up to 500 meters with minimal attenuation. This type of cable is often used as a backbone cable. A transceiver is used to connect thinnet COAX to the larger thicknet COAX. There are two types of grade associated with COAX cables—polyvinyl cable (PVC) or plenum. If a PVC cable is burnt, it will give off a poisonous gas. Plenum cable, on the other hand, is fire resistant, and produces only a small amount of smoke.

The advantage of a COAX cable is that it has very low EMI sensitivity and a high bandwidth, but the disadvantage is that the cost is much more expensive than UTP or STP cables. Below is a summary list of COAX cable types and their descriptions:

Cable Type	Description
RG-58 /U	Solid copper core, uses 50-ohm impedance.
RG-58 A/U	Stranded wire core, uses thinnet and 50-ohm impedance.
RG-59	Cable TV uses 75-ohm impedance.
RG-6	Larger in diameter and rated with higher frequencies than RG-59, can be used for broadcast transmission as well.
RG-62	Used for ARCNet network. It uses 93-ohm impedance.

Fiber optic cables are cables that transmit data through a thin glass or plastic fiber using light waves. They can carry digital data at 4 Gbps. Fiber optic cable is the safest way to transmit data, because no electrical impulses are carried with the signal. It has a very high speed and can carry a large amount of information. Fiber optic cable has an extremely thin cylinder of glass, called the core, which is surrounded by a concentric layer of glass called cladding. Each glass strand passes signals in only one direction, and so a fiber optic cable consists of two strands in separate jackets. Fiber optic cables are not subject to any electrical interference and are extremely fast. Thus the signal cannot be tapped easily. Signals can travel up to hundreds of kilometers without significant attenuation. The advantages of a fiber optic cable are that it has a very high bandwidth and no sensitivity to EMI. The disadvantage is that it is the most expensive type of bounded media. Below is a list of factors for each of the four types of cables:

Factor	UTP	STP	COAX	Fiber Optic
Cost	Lowest	Moderate	Moderate	Highest
Capacity	1 to 100 Mbps. Typically 10 Mbps	1 to 155 Mbps. Typically 16 Mbps	Typically 10 Mbps	Up to 2 Gbps. Typically 100 Mbps
Attenuation	High, in the range of a hundred meters	High, in the range of a hundred meters	Low, in the range of a few km	Lowest, in the range of tens of km
EMI	Most sensitive to EMI and eavesdropping	Less sensitive than UTP, but sensitive to EMI and eavesdropping	Same as STP	Not affected by EMI or eavesdropping

Table 6.1 Cable specifications

10base 2 (8) Coax (thin)
10base 5 Thick Coax cable
10base T Twisted pair

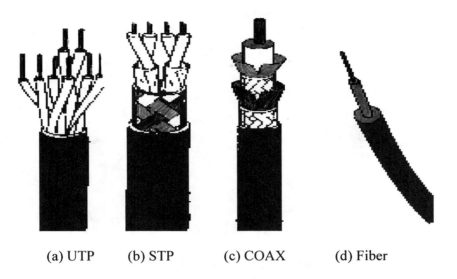

(a) UTP   (b) STP   (c) COAX   (d) Fiber

Figure 6.4 Different types of cables

If you have an extended local area network, you might need multipoint wireless connectivity, which is a wireless bridge that links two buildings without using cables. In addition, if the buildings are 25 or more miles apart, then a spread-spectrum radio transmission can be used. This will eliminate the need for a T1 line or a microwave connection. A T1 line transmission carries data and voice at the rate of 1.544 Mbps.

If you have a mobile computing network, you might have to use a telephone line to transmit and receive data and voice. The telephone line transmission rate range is approximately 110 Kbps to 56 Kbps. In addition to telephone line, you might want to use a cellular network, which is faster, but suffers from subseconds delays.

For a wireless communication (unbounded media) the transceiver (access point) broadcasts and receives signals to and from the surrounding computers. It uses a small wall-mounted transceiver to the wired network, and it establishes radio contact with portable networked devices. Now let us look at four types of unbounded media: radio, microwaves, laser, and infrared. Radio waves are electromagnetic waves in the frequency range of kilohertz to low gigahertz. Some radio waves have broad beams and high EMI sensitivity. A PC tunes to narrow-band (single frequency) radio waves (like radio broadcasting). It can transmit up to 5000 square meters, but cannot transmit through steel or load-bearing walls. It can transmit data at 4.8 Mbps. Spread-spectrum radio waves are waves that broadcast over a range of frequencies. These frequencies are divided into channels or hops. They can transmit data up to 250 Kbps.

Microwaves are electromagnetic waves that fall near light waves in frequency. Microwaves have to transmit by a "line of sight" and cannot transmit around the world. Microwaves make use of terrestrial relay stations to transmit data for a few miles. If a microwave signal needs to be transmitted over a longer distance, the satellite communication link that is orbiting the earth will be used. Microwaves have moderate EMI sensitivity and a narrow or broad beam.

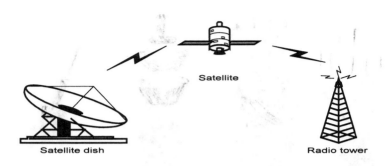

Figure 6.5 Microwave communication

Laser is a light beam operating at one frequency. The waves of a laser are synchronized, and are transmitted in one direction. It is the most ideal type of unbounded media, because it has no EMI sensitivity due to a very narrow beam. Infrared transmission transmits several narrow beams at once from one station to another. It can transmit up to 1 Mbps as long as the distance is within 10 feet. It does not have a strong ambient light. Infrared line-of-sight transmission must be clear, while scattered infrared transmission limits the transmission to 4 Mbps at 3 feet, because the signals bouncing off walls and ceilings attenuate the signal. In a reflective infrared transmission, optical signals are transmitted toward a common location, where they are redirected to the appropriate PC. A broadband optical telepoint infrared transmission can handle the high quality requirements of multimedia, which in turn can provide broadband services such as bounded media. This technology is used mostly in television and wireless networks.

Transmission Devices

There are two types of transmission devices: communication devices and interconnection devices. Communication devices include modems, codecs, and MUXs. Interconnection devices include repeaters, hubs, bridges, routers, and gateways.

One of the most popular communication transmission devices is the modem. A modem is a hardware device that lets you connect two computers using standard phone lines. First, the sending computer's modem **mo**dulates the computer's digital signals into analog signals that can pass over the phone lines. Then the receiving computer's modem **dem**odulates the analog signal back into the digital signal that computers understand. There are two types of modems—internal and external. Internal modems are based on a board, which can be plugged into any of the expansion slots in a PC. External modems are plugged into the RS232 port or the COM port of a PC with a RS232 cable. The speed of a modem is measured in baud—the speed of the oscillation of the sound wave on which a bit of data is carried over the telephone wire. The bps (bits per second) can be greater than the baud rate due to encoded data, so that each modulation of sound can carry more than one bit of data over the telephone line.

Often you will encounter the terms DTE (Data Terminal Equipment) and DCE (Data Communication Equipment). A modem is a type of DCE that interfaces with a DTE. On the other hand, a codec (COder/DECoder) is a hardware device that converts analog data for transmission on a digital medium. A codec actually functions as a mirror image of a modem. A multiplexer (MUX) is a hardware device that allows multiple simultaneous signals to share a single transmission medium.

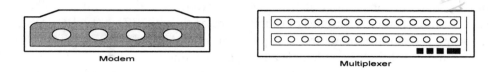

Figure 6.6 A modem and a multiplexer

Interconnection devices are devices that connect computers or networks together. A repeater is an interconnection device that operates at the physical layer. Its main purpose is to amplify or reshape a signal to extend the range of the network. Repeaters do not translate or filter anything. Both segments on a repeater must use the same access method (i.e., CSMA/CD). Repeaters can take an Ethernet packet from a thinnet COAX and pass it onto a fiber optic segment.

A hub, sometimes called a concentrator, is a device that connects workstations to a central point (hub). It is a special kind of repeater and/or bridge designed to facilitate a star topology. There are three types of hubs: passive, active, and intelligent. A passive hub splits the signal and only routes traffic to all nodes. Active hubs perform similarly to passive hubs, but they regenerate, amplify, and retransmit signals. Intelligent hubs perform similarly to passive and active hubs, but they route traffic only to the branch of the receiving node and perform some network management. You can think of bridges and routers as intelligent hubs.

A bridge is an intelligent device used to connect two separate network segments. A bridge uses the physical addresses of the source and destination to separate and keep traffic to a minimum on each segment of the network. A bridge has all the features of a repeater. It works at the MAC (Media Access Control) sublayer. As traffic passes through bridges, information about the PC's addresses is stored in the bridge's RAM. This RAM builds a bridging table based on the source addresses. If a bridge knows the location of the destination node, it forwards the packet to it. If it does not know the destination, it forwards the packet to all the segments. A bridge regenerates the data at the packet level, and packets can travel by more than one route. Since a bridge has access to the physical address, it operates at the data link layer.

A router is a device that makes use of the logical address of the network. A router can minimize traffic by connecting two networks together and keeping each network segment's traffic on its own side of the network. A router determines the best path for sending data, and it filters broadcast traffic to the local segment. A router filters and isolates traffic, connects network segments together, and passes information only if the network address is known. The routing table in the router contains all the known network addresses, how to connect to other networks, all the possible paths between those routers, costs of sending data over those paths, and the media access control layer addresses, as well as the network addresses (logical addresses). A router can choose the best path available. It decides the path the data packet will follow by determining the number of hops between the internetwork segments.

There are two types of routers—dynamic and static. Dynamic routers are automatic discovery routers, while static routers are manually configured to set up a routing table to specify each route. A router not only recognizes an address, as a bridge does, but also recognizes the type of protocol. In addition, a router can identify addresses of other routers, and determines which packet to forward to which router. Examples of routable protocols are DECnet, IP, IPX, OSI, XNS, and DDP (Apple). Examples of nonroutable protocols are LAT and NetBEUI. TCP/IP is supported by the protocol called OSPF, which is a link-state routing algorithm. IPX is supported by the protocol called NLSP, which is a link-state algorithm. In addition, IPX and TCP/IP are supported by the protocol called RIP, which uses a distance-vector algorithm to determine the routes. Since the router makes use of the logical address, it operates at the network layer.

Brouters are bridges and routers combined in one device. Remember that a bridge does not support routable protocols. A brouter can operate at either the data link or the network layer.

Gateways are devices that connect two completely different types of network together. A gateway repackages and converts data from one environment to another, so that one environment can understand the other environment's data. Gateways link two systems that do not have the same protocols, data formatting structure, languages, and architecture. Gateways operate at the transport layer and at higher layers of the OSI model.

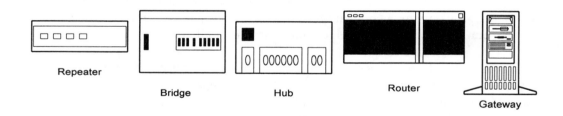

Figure 6.7 Different types of interconnection devices.

Network Topologies

Network structures are often referred to as network physical topologies. The physical topology of a network is the layout of the cabling, or how computers are connected in a network. There are two types of physical topologies: point-to-point and multipoint. A point-to-point (PTP) physical topology is the connection of two nodes directly together.

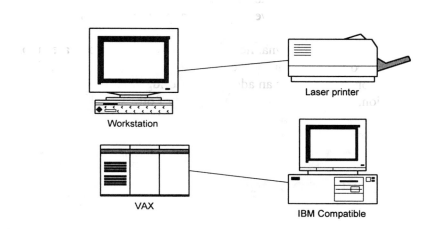

Figure 6.8 Examples of PTP topologies

Multipoint topology is the connection of three or more nodes together. There are six different types of multipoint topologies: peer-to-peer, bus, star, ring, mesh, and hybrid.

A peer-to-peer topology is the connection of more than two nodes directly together. You should set up a peer-to-peer network if you have fewer than 10 nodes, the network is located in one location, and security is not an issue. The rest of the following multipoint topologies are mostly server-based topologies, where security is often an issue.

In a bus topology, all nodes are connected to the same transmission medium, usually an electrical cable. The end of the cable must be terminated with a T terminator to prevent data reflection from coming across and causing data corruption. The linear bus is connected with a single cable. It communicates by addressing data to a particular computer. When sending data, it uses a passive topology, which means that only one PC can send messages at a time using address matching. To expand the network, you may use a barrel connector to connect two cables together to make a longer one, or use a repeater to boost the signal and send it on its way. The advantage of a bus topology is that the cabling cost is minimal. The disadvantage is that it is very hard to troubleshoot. If the cable breaks, it can disable the whole network.

In a star topology, all nodes connect to a central point. The central point could be a hub or a concentrator. The advantages of a star topology is that each device has its own cable to connect to the central point, thus making it easier to disconnect the device without disabling the network. In addition, it is easy to expand and many different types of cables can be used; and it is a centralized monitoring of the network activities and traffic. Because troubleshooting is easy in a star topology, it is the most popular method of connecting computers. The disadvantage of a star topology is that it has a single point of failure (hub or concentrator). In addition, more cabling is required in a star topology, since each node must be connected to the central point.

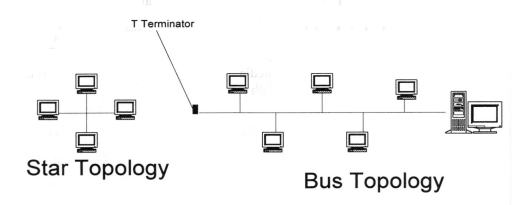

Figure 6.9 A star and a bus topology

In a ring topology, the nodes are connected in a closed loop. Each node on the ring receives and regenerates the signal before it retransmits it. Therefore, the advantage of a ring topology is that each node acts as a repeater. The disadvantage of a ring topology is that if the ring breaks, it will disable the whole network. Also, the networking devices for a ring topology are expensive.

In a mesh topology, each node must be connected with at least two other nodes. As you can probably see, a mesh topology is not practical in a large network. The advantage of a mesh topology is that multiple links exist between nodes. You can always find a path to route signals in case of a node failure. The disadvantage of a mesh topology is that a lot of networking devices are needed, such as cables and NICs, thus increasing the cost.

Finally, a hybrid topology is a mixture of two or more topologies in one network. The advantage of a hybrid topology is that an organization can use it to customize workgroup efficiency and traffic. The disadvantage is that devices from one topology cannot be placed into another topology without hardware changes.

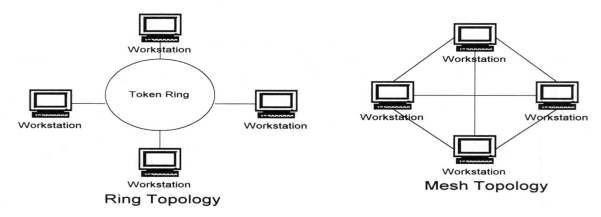

Figure 6.10 A ring and a mesh topology

Table 6.2 shows a summary of the advantages and disadvantages of multipoint topologies:

Topology	Advantages	Disadvantages
BUS	1) Cable is cheap 2) Media is easy to work with 3) Simple, reliable 4) Easy to expand	1) Slow in heavy traffic 2) Problems are hard to isolate 3) Cable break can affect the network
STAR	1) Easy to expand and modify 2) Centralized monitoring and management 3) One node failure does not affect the network	1) If the centralized point goes down, the whole network goes down 2) Slow unless switched
RING	1) Equal access to all PCs 2) Even performance despite many users	1) One PC going down affects the network 2) Problems are hard to isolate 3) Hardware is expensive
MESH	Signal can always find a path despite node failures	Cost from many networking devices
HYBRID	Customized workgroup efficiency and traffic	Cannot be placed with other topologies without hardware changes

Table 6.2 Advantages and disadvantages of multipoint topologies

Although physical topologies have the same names as logical topologies, logical topologies describe the way data are traveling on the network. There are two types of logical topology associated with networks—ring and bus. It is possible to have different physical and logical topologies in the same network. In a ring logical topology the data are passed from one node to the next. With a bus logical topology, the data are passed to all the nodes on the network at once, and data will always be available to all nodes on the network.

The media access method (sometimes referred to as logical topology), is subdivided into three types: contention, token passing, and polling. On a contention network, any device can transmit data whenever it needs. Since devices can transmit data at any time, data collisions will occur frequently. To avoid data collisions, a specific protocol was developed. The Carrier Sense Multiple Access with Collision Detection (CSMA/CD) was developed to avoid data collision on a network having contention as its media access method. CSMA/CD works by "listening" to the cable for traffic while transmitting information. If collision occurs, data will be retransmitted. CSMA/CD is the access protocol used in Ethernet and IEEE 802.3.

CSMA/CD is not effective beyond 2,500 meters. AppleTalk networks use the CSMA/CA (Carrier Sense Multiple Access with Collision Avoidance) protocol as its main access mechanism. The advantages of contention are that it is a very simple access method, and that it provides high data throughput at low traffic. The disadvantage of contention is that it is probabilistic and not deterministic. In addition, at high traffic levels, data collisions occur more frequently, resulting in poor performance.

On a token passing network, a token is passed to one device at a time in an orderly manner (that is why a token ring network is considered to be deterministic rather than probabilistic), and only devices that have the token may transmit data. The advantage of token passing is that it offers high throughput at high traffic. The disadvantages of a token passing network are that its protocols for managing the network are complicated, and that devices using token passing are much more expensive. The token passing method is used with networks such as IEEE 802.4 (Token Bus topology), IEEE 802.5 (Token Ring topology), ARCnet, and TokenTalk for Apple's Macintosh.

RESTRICTIONS	VALUE/DESCRIPTION
Network	ARCnet
Physical topology	Star or bus
Logical topology	Bus
Access control method	Token passing
Cables	COAX (common), UTP or fiber
Data speed	2.5 Mbps, broadband
Maximum distance between nodes	2,000 feet (610 meters), star topology
Maximum distance between two farthest nodes	20,000 feet, star topology
Maximum distance between node and a passive hub	100 feet, star topology
Maximum distance between node and an active hub	2,000 feet, star topology
Maximum distance between passive hub and an active hub	100 feet, star topology
Maximum distance between two active hubs	2,000 feet, star topology
Using 93-ohm RG-62 A/U	610 meters maximum, in a star topology; will be 305 meters maximum in a bus topology
Using RJ-11, RJ-45 UTP	244 meters maximum on a star or a bus topology

Table 6.3 Restrictions on the ARCnet network

RESTRICTIONS	VALUE/DESCRIPTION
Network	Ethernet 10Base2
Physical topology	Bus
Logical topology	Bus
Access control method	Contention-CSMA/CD
Cables	Thin COAX
Data speed	10 Mbps, baseband
Minimum distance between workstations	½ meter (1 ½ feet)
Maximum segment length	185 meters (607 feet)
Maximum network length	925 meters (3,035 feet)
Maximum node separations	5 segments/4 repeaters
Maximum nodes per segment	30
Maximum populated segments per LAN	3
Connectors	BNC barrel, BNC T-connectors, or BNC terminators
5-4-3 rule applies	5 cable segments are connected with 4 repeaters, but only 3 segments can have stations attached

Table 6.4 Restrictions on the Ethernet 10Base2 network

RESTRICTIONS	VALUE/DESCRIPTION
Network	Ethernet 10Base5
Physical topology	Bus
Logical topology	Bus
Access control method	Contention-CSMA/CD
Cables	Thick COAX
Data speed	10 Mbps, baseband
Minimum distance between transceivers	2 ½ meters (8 feet)
Maximum transceiver cable length	50 meters (164 feet)
Maximum segment length	500 meters (1,640 feet)
Maximum network length	2,500 meters (3,035 feet)
Maximum node separations	5 segments/4 repeaters
Maximum taps per segment	100
Maximum populated segments per LAN	3
Connectors	DIX , AUI, N-series connectors or terminators
5-4-3 rule applies	5 cable segments are connected with 4 repeaters, but only 3 segments can have stations attached

Table 6.5 Restrictions on the Ethernet 10Base5 network

RESTRICTIONS	VALUE/DESCRIPTION
Network	Ethernet 10Base-T
Physical topology	Star
Logical topology	Star
Access control method	Contention-CSMA/CD
Cables	UTP with RJ-45 connectors, STP, category 3,4,5
Data speed	10 Mbps, baseband
Maximum distance between workstation and hub	100 meters (328 feet)
Maximum nodes per segment	512
Maximum node separations	5 segments/4 repeaters
Maximum hubs in sequence	4
Maximum populated segments per LAN	3

Table 6.6 Restrictions on the Ethernet 10Base-T network

In addition to the above networks, there are other Ethernet networks in use today. These include 10BaseFL, 100VG-AnyLAN, and 10BaseX. 10BaseFL networks operate at 10Mbps (baseband), use fiber optic cable, and have a maximum segment length of 2000 meters. In the network 100VG-AnyLAN, (VG stands for Voice Grade) data is transmitted at 100 Mbps, and is connected with category 5 UTP, STP cables, or a fiber optic cable. It uses polling as its access control method, and can support both Ethernet and token ring packets. It has a star topology with child hubs that act as computers to their parent hubs. In a 100BaseT hub, links cannot exceed 250 meters. 100BaseX is a fast Ethernet that uses a UTP category 5 cable. It uses CSMA/CD as its access control method. 100BaseT4 networks use 4-pair category 5 UTP or STP cable. 100BaseTX networks use 2-pair category UTP or STP. 100BaseFX networks use 2-strand fiber optic cable.

In a token ring network frame, the media access control field indicates whether the frame is a token or a data frame. When the first token ring PC comes online, the network generates a token. Only one PC, which has the token, can send data over the network. After it finishes sending data, the PC removes the frame from the ring and transmits a new token back on the ring. In a token ring network, each PC acts as a unidirectional repeater.

RESTRICTIONS	VALUE/DESCRIPTION
Network	Token ring
Physical topology	Star
Logical topology	Ring
Access control method	Token passing
Cables	UTP or STP with UDC connector
Data speed	4 or 16 Mbps, baseband
Maximum number of MAUs (Multi-Access Units) that can connect to each other	12 using STP
Maximum number of nodes	72 using UTP
Maximum number of nodes	96 using STP
Maximum patch cable distance connecting all MAUs	400 feet
Maximum patch cable distance between two MAUs	150 feet
Maximum distance between MAU and a node	150 feet
MAU to PC for IBM type 1 cable	101 meters
MAU to PC for IBM type 3 cable	150 meters
MAU to PC for UTP cable	45 meters
MAU to PC for STP	100 meters

Table 6.7 Restrictions on the Token Ring network

RESTRICTIONS	VALUE/DESCRIPTION/MATERIAL NEEDED
Network	FDDI
Physical topology	Ring or star
Logical topology	Ring
Access control method	Token passing
Cables	Fiber
Data speed	100 Mbps
Maximum number of nodes in a ring topology	500
Maximum total cable length in a ring topology	100,000 meters

Table 6.8 Restrictions on the FDDI network

Most data is routed through a network using two routing methods, route discovery and route selection. Route discovery is the process of finding all possible routes through the internetwork. Route discovery then builds a routing table to store that information. There are two methods of route discovery: distance-vector and link-state. In distance-vector routing, each router on the network periodically broadcasts the information in the routing table. The other routers then update their routing tables with the broadcast information they receive. NetWare 4.11 uses the concept of distance-vector routing. In link-state routing, routers broadcast their complete routing tables only at startup, and at many fewer time intervals, to minimize network traffic.

Route selection is the process of answering the question, "What is the best way to get there." There are two types of route selections: dynamic and static. Dynamic route selection uses the cost information in routing tables to find the best way through the network. Cost information (sometimes measured as hops) indicates the different paths the network has used to transfer data from one place to another. The router has the capability to select new paths "on the fly" as it transmits data. In static route selection, the data are designated in advance. They are not selected from routing tables or programmed by an administrator.

Data are divided into smaller pieces of information called packets. The packets are transmitted through the network using the method of packet switching. Packet switching uses both circuit and message switching to avoid the disadvantages of both. In packet switching, information is broken into packets. Each packet contains a header with source, destination, and intermediate node address information. The difference between packet switching and message switching is that packet switching limits the length of a packet. Since packets are much smaller than messages, packets can be stored and forwarded from memory, resulting in a much faster and more efficient way of transferring data. Packet switching uses a virtual circuit—a logical connection between the sending computer and the receiving computer. Data paths for the individual packets depend on the best route at any given instant.

There are two types of packet switching: datagram and virtual-circuit. In datagram packet switching, each packet is treated as if it were a whole message rather than a piece of something larger. Virtual-circuit switching establishes a logical connection between the sender and the receiver devices. All packets in virtual-circuit switching will flow in this logical path.

WAN Protocols

This section will primarily cover miscellaneous protocols used in WAN networks. For example, in WAN networks digital signals are used in Digital Data Services (DDS). DDS provides point-to-point synchronous communication at 2.4, 4.8, 9.6, or 56 Kbps. DDS can be made into a guaranteed full-duplex bandwidth by setting up a permanent link from each end point. It is 99% error-free and does not require a modem.

Furthermore, in a WAN network, digital signals can be transmitted on a T1, T3, or a switched 56 line. A T1 line is point-to-point transmission and uses two wire pairs (one pair to send, one pair to receive). It is a full-duplex signal that transmits at 1.544 Mbps. It uses multiplexing, which involves signals from different sources that are collected into a component called a multiplexer, and then fed into one cable at the speed of 8,000 times a second. Each channel can transmit at 64 Kbps (DS-0) or at 1.544 Mbps (DS-1). T3 line can transmit at 6 Mbps to 45 Mbps. Conversely, a switch 56 line can transmit at 56 Kbps, but it can only be used on demand and requires a CSU/DSU to dial up another switched 56 sites.

The IEEE (Institute of Electrical and Electronics Engineers) is responsible for the 802 series standards. The 802 series standards are important in the world of LAN networks, since they define Ethernet and token ring networks. The following table (Table 6.9) defines the IEEE series, the standard it defines, and the layer of the OSI model to which it corresponds.

IEEE	STANDARD	OSI MODEL
802.1	Overall	Physical to transport
802.2	LLC	Logic control layer
802.3	Ethernet-CSMA/CD	Physical and data link
802.4	Token bus-Token passing	Physical and data link
802.5	Token ring-Token passing	Physical and data link
802.6	MAN network	Physical and data link
802.7	Broadband technology	"Under development," physical and data link
802.8	FDDI-token passing	Physical and data link
802.9	Integrated voice and data (ISDN)	LLC
802.10	LAN Security	Network to session
802.11	Wireless LANs	"Under development," physical and data link
802.12	100VG-AnyLAN	"Under development," physical and data link

Table 6.9 IEEE 802 series

Other WAN protocols include SLIP, PPP, FDDI, X.25, Frame relay, ISDN, SONET, ATM, and SMDS.

SLIP (Serial Line Internet Protocol) and PPP (Point-to-Point Protocol) are used with dial-up connection to the Internet. PPP is an improved version of SLIP.

FDDI (Fiber Distributed Data Interface) is a LAN or a MAN standard that can be physically connected in a ring or a star. It uses a token passing media access method similar to IEEE 802.5. It is a 100 Mbps token ring network that uses fiber optic cables. Its maximum ring length is 100 km and 500 PCs. It is different from the IEEE 802.5 standard. FDDI can transmit as many frames as it can produce within a predetermined time before letting the token go. Furthermore, more than one computer can transmit at the same time with an FDDI network. The computer detects a fault and then sends a "beacon" to its upstream neighbor, and so on, until it reaches the computer that originally sent the beacon. The original computer then receives its own beacon back and assumes that the problem is fixed; then it regenerates the token.

The CCITT (Consultative Committee International Telegraph and Telephone) developed the X.25 protocol. X.25 is a WAN standard protocol that uses packet switching technology. It uses switches, circuits, and routes as available to provide the best routing to any particular time. It also uses telephone lines, which are slow, and has error checking. Finally, the X.25 protocol requires a DTC/DCE interface, and it can be a synchronous packer-mode host, as in a Public Data Network (PDN over a dedicated leased-line circuit, or some other device).

Frame relay is similar to X.25 protocol. It makes use of packet switching technology with virtual circuits. Frame relay is a fast packet, variable-length, and digital technology. In addition, it requires a frame-relay-capable router or a bridge. It is a point-to-point technology, and uses PVC to transmit at the data link layer over a leased digital line.

ISDN (Integrated Services Digital Networks) provides transmission of voice, video, and data over digital telephone lines. It uses 3 data channels—2 for 64 Kbps, and 1 for 16 Kbps. The 64 Kbps channels are known as B channels and carry voice, data, or images. The 16-Kbps channels are known as the D channels which carry signaling and link management data. The basic rate = 2B + D. Finally, ISDN is a dial-up service, not dedicated, and not bandwidth on demand.

SONET (Synchronous Optical Network) is a physical layer protocol generally used by WAN networks. It is a synchronous optical network that uses fiber optic cables. It works with frequencies larger than 1 Gbps.

ATM (Asynchronous Transfer Mode) is a data link layer mostly used for WANs. It uses cell-switching technology. A cell is a 53-byte packet. It is a fixed-sized packet over broadband and baseband LANs or WANs. It can be transmitted at 155 Mbps, 622 Mbps, or more. It is a broadband cell relay method that transmits data in 53-byte cells (each with 48 bytes of application information) rather than in variable-length frames. You can transmit up to 1.2 Gbps or more. An ATM requires hardware, such as routers or bridges, that is compatible with ATM. ATM also requires switches (multiple hubs; router-like devices). The transmission rate of ATM depends on what media are being used:

Media	Rate
FDDI	100 Mbps
Fiber channel	155 Mbps
OC3 SONET	155 Mbps
T3	45 Mbps

SMDS (Switched Megabit Data Service) is similar to ATM. It also uses cell switching, but it can be mapped into the data link and network layers of the OSI model. It is a multimegabit data service that can transmit at 1 Mbps to 34 Mbps.

Questions

1) What is the purpose of the OSI model?
2) List all the layers of the OSI model, and name the data in each layer.
3) Describe the function of each layer of the OSI mode.
4) Where do the following devices operate in the OSI model? Describe the function of each.

   Gateway, Repeater, Bridge, Router

5) What is the transmission media for a 10BaseT?
6) What are the possible physical topologies and media access control methods for the following networks?

   FDDI, 10Base5, Token Ring, 10BaseT, 10Base2

7) What is the name of the node that ensures the integrity of the token as it travels along the channel?
8) Describe each of the following:

     a) Centralized computing
     b) Distributed computing
     c) Collaborative computing
     d) Network models
     e) MAN (Metropolitan Area Network)
     f) Network services
     g) Fiber optic cable
     h) COAX cable
     i) Active hub

9) What are the advantages of having a bounded media?
10) What is the capacity of a UTP?
11) Which bounded media has the worst attenuation and EMI?
12) What type of cable would you use to connect 100 PCs on one floor of a building?
13) Name all of the communication devices and interconnectivity devices.
14) Name all the most popular multipoint physical topologies and name the two main components required in a physical topology.
15) List all the advantages and disadvantages of the most popular multipoint physical topologies.
16) What does beaconing mean?
17) Describe the following:

     a) TDM
     b) StatTDM
     c) FDM

18) Describe the following:

     a) Packet switching
     b) Message switching
     c) Circuit switching

19) Define all 12 IEEE 802 series standards. How does each compare to the OSI model?
20) Name an emerging broadband ISDN standard that uses cell relay technology to perform network layer activities.

7

The Internet

This chapter summarizes the concepts of the Internet. It will cover the four layers of the DOD model (TCP/IP model) and each of the layer protocols. In addition you will learn to distinguish among HTTP, WWW, HTML, and URL terms. In addition you will learn IP addressing and how to create a web page using HTML.

Introduction to TCP/IP and the Internet

The Internet is a worldwide network of networks. TCP/IP (Transmission Control Protocol/Internet Protocol) is the communication protocol for the Internet and was introduced by an experiment conducted by the DOD (Department Of Defense) and several universities in the mid 1970's. DOD was looking for a way to connect all of their computer systems using existing resources. To have an internetwork of the many devices currently in place, the individual networks needed to be connected. The devices that connected these individual networks were called gateways, and a packet-switched network was the specific implementation that was chosen to exchange data.

TCP/IP is a set of common networking protocols used by computers for exchanging data. DOD defines TCP/IP as a four-layer networking model. Each layer of the TCP/IP model consists of a number of protocols. These protocols are collectively referred to as the TCP/IP protocol suite. TCP/IP was developed to interconnect computers and networks with widely different operating systems throughout the world on the Internet. It offers users immediate access to a complete spectrum of services and information. It has evolved into today's Internet connecting thousands of computers and networks throughout the world.

TCP/IP is the standard networking protocol for UNIX systems. It comes as a built-in part of many versions of UNIX. It is a standard part of 4.2 and 4.3 BSD UNIX. It has been added to many AT&T UNIX System V implementations as well, and it has become synonymous with UNIX networking. Specifications have been written for a complete set of TCP/IP applications for file sharing, terminal emulation, file transfer, printing, messaging, and network management. The specifications are freely available to any developer wanting to write an application. Today you can find commercial, shareware, and public domain versions of TCP/IP for all types of computers including PCs, SUNs, VAXs, Macintoshes, minicomputers, and mainframes. More than 100 vendors offer products supporting TCP/IP and related protocols.

In summary, the combination of TCP/IP networks together with an open-door policy that allowed commercial research networks and academic facilities to connect to the DOD generated the "super information highway" called the Internet. Because of the need for convenient and reliable communications a set of protocols was needed. Out of this need came the TCP/IP protocol suite.

When you connect to the Internet, you can access thousands of computers throughout the world. Below are the most popular services available on the Internet:

- WWW (World Wide Web)—WWW is a collection of approximately one billion linked hypertext documents stored on thousands of servers on the Internet. It is also a type of data service running on many computers for easily traveling (surfing) across the Internet. It uses a hypertext to associate text with a URL (Uniform Resource Locator) that can point anywhere on the Internet. A URL identifies a document or a location on the Internet. For example, the URL for the America Online home page is http://www.aol.com. You can generally type the URL into any browser (e.g., Netscape) and the browser will connect you to the specified location.

  Using browser programs such as Netscape or MS Explorer, you may access WWW servers to obtain the latest information. Other types of documents that are available through WWW are Gopher Text, multimedia, and FTP files.

- Gopher—Gopher is another tool on the Internet. It was created at the University of Minnesota, where the school mascot is the gopher. It organizes topics into a menu system and allows you to access the information on each topic listed. The menu system includes many levels of submenus, allowing you to burrow down to the exact type of information you are looking for. Gopher actually uses the Telnet protocol to log the user in to other systems. Gopher servers act as master directories of the Internet. Those directories can provide the specific location of a file or topic.

- Telnet—Telnet is a protocol that you may use to log on to any host attached to the Internet by providing an account name and password for the host. You have complete access to all the resources on the computer, but you must know how to operate the computer you signed onto.

- FTP (File Transfer Protocol)—FTP is a protocol that allows you to transfer files between any two computers that are using FTP. Hundreds of public domain, freeware, and shareware programs are available on anonymous FTP servers located throughout the Internet. These FTP servers are called anonymous because anyone can log on to the server using an "anonymous" account name and e-mail address as the password.

- USEnet—USEnet is a very large bulletin board system made up of thousands of different conferences. It currently offers over 8,000 bulletin boards and forums. Forum members may exchange data on any topic.

- E-mail—E-mail is when a user sends an electronic mail to any user connected to the Internet. Messages may contain text, graphics, voice messages, or application databases.

As you can see from the list above, Internet users may access resources from USEnet, FTP servers, Gopher servers, and WWW. Now let us answer the question "How are computer able to talk to each other despite the distance and diversity?" The answer is a common set of protocols, which enables all TCP/IP hosts to talk to each other. These protocols have specific formats and conventions that all TCP/IP hosts follow. There are four layers of protocols that make up the TCP/IP or the DOD model. Each layer is responsible for performing specific networking functions. Each layer may include several protocols. To understand TCP/IP, it is important to examine the function of each layer and identify the individual protocols that collectively make up TCP/IP. The four layers of the DOD model are as follows:

Layer 4: Process/Application
Layer 3: Host-to-Host
Layer 2: Internet
Layer 1: Network Access

Here is the best way to remember the order of the layers:

Statement	DOD layers
Put All	Process/Application
Hardware	Host-to-Host
Into	Internet
Networking	Network Access

The DOD model compares with the OSI model in many aspects with some exceptions. Below is a chart to help you remember the OSI model layers and their primary functions:

Layer Number	Layer Name	Layer Function
7	Application	User interface. Command interfaces with network.
6	Presentation	Data translation. Data is converted to and from computer code.
5	Session	Dialogue management. Sender and receiver connections are set up.
4	Transport	Reliability. Quality and reliability of data transmission is managed.
3	Network	Routing. Data is routed through the network.
2	Data-Link	Packaging. Data is packaged for transmission.
1	Physical	Real connection. Bits and bytes are transferred from the sender to the receiver.

The chart below shows how the OSI model and the DOD model compare:

DOD Model	OSI Model
Process/Application	Application
	Presentation
	Session
Host-to-Host	Transport
Internet	Network
Network Access	Data-Link
	Physical

The specifications for each of the protocols within the TCP/IP suite are defined within one or more RFC's (Request For Comments). Various users on the Internet submit these RFCs for proposing new protocols, suggesting improvements of existing protocols, and/or offering comments on the state of the network. You may obtain electronic copies of all RFCs from the anonymous FTP server at DS.INTERNET.NET or other servers throughout the Internet. Here is a list of the most commonly used protocols and the layers at which they function.

Layer 4: Process/Application Protocols

This layer is responsible for the user's interface with the network. TCP/IP applications usually include a client and a server program. A user executes the client program when access to the server is desired. The server program is often referred to as a "daemon." *Daemon* comes from Greek mythology and means a "guardian angel." Usually a daemon starts when the system boots up. The program runs in the background on the host server. Typically, a daemon process is not continually running on the server. Some sort of event, such as a client request, triggers it. The Process/Application layer has a set of standardized protocols to provide terminal emulation, file transfer, electronic mail, and other applications such as:

1) FTP (File Transfer Protocol)—You can use FTP to transfer files between a PC and a remote host. In addition, you may use FTP to access directories and files on the remote system and list, copy, and manage directories and files on the remote system. Files can be transferred in ASCII or binary, depending on whether you want to transfer text or programs. When you use FTP, you will need an account login ID and a password.

2) TFTP (Trivial File Transfer Protocol)—This protocol is similar to FTP. TFTP does not support an account name or password and takes fewer overheads than FTP because FTP uses TCP, while TFTP uses UDP. TCP and UDP protocols are discussed in the next section. TFTP may be used to download, transfer, or copy files between two systems without specifying an account or a password. However, you may not use TFTP to list files or directories. Since TFTP does not use accounts and passwords, most TFTP implementations restrict the types of files a workstation can access and deny access to a file unless every user on the host can access the file.

3) TELNET (Virtual Terminal Emulation)—This protocol provides access to a computer connected to the network. The connection is in the form of a terminal session that appears to users to be hard-wired directly to the host. The interface looks the same as it would if you were using the console on the host itself. A TELNET application lets you emulate a terminal connected to a remote host. Once you log into the remote host, you can execute commands and perform any operations the remote host supports.

4) SMTP (Simple Mail Transfer Protocol)—This protocol is used to communicate with other network users. You may use SMTP to send and receive electronic mail (e-mail). SMTP is the engine that delivers mail on the TCP/IP host.

5) NFS (Network File System)—This protocol lets your file server provide services as a distributed file system for TCP/IP hosts.

6) Broadcasts—This protocol is a packet delivery system that delivers a copy of a given packet to all attached hosts.

7) LPD (Line Printer Daemon)—This protocol is designed for printer sharing. It allows print jobs to be spooled and sent to network printers.

8) RPR (Remote Printing)—This protocol is designed for remote printing.

9) SNMP (Simple Network Management protocol)—This protocol is designed for network management.

10) X Windows—This protocol is for application sharing. It is designed for client-server operations. It defines a protocol for writing of a graphical user interface based on a client-server application.

Layer 3: Host-to-Host Protocols

This layer is responsible for creating and maintaining connections between communicating hosts. It is responsible for the following:

a) The integrity of data transfer.
b) Setting up reliable, end-to-end communications between systems.
c) Providing error-free delivery of data units, in proper sequence, and with no loss or duplication.
d) Providing reliable, sequenced delivery of messages.

There are two protocols that function at this layer:

1) TCP (Transmission Control Protocol)—This protocol provides a reliable connection between communicating hosts. It also sequences and acknowledges packets. It establishes a virtual circuit for communications. Upper-level protocols that use TCP protocol include FTP, SMTP, and TELNET. There is more overhead because of the tracking that must be done to ensure that the packet is transmitted smoothly with no error.

   TCP is like a polite telephone conversation. Before you can speak with someone on the phone, you dial his or her number and establish connection with the other person. This is like a virtual circuit with the TCP protocol. During the conversation, you may periodically ask, "Did you get this?" This is like TCP acknowledgement. You may also ask during your conversation "Are you still there?" to verify that the phone connection still exists. At the end, you may end the conversation and hang up the phone.

   TCP connection begins with a client requesting a virtual connection from a remote host. No communication is possible until the remote host responds. Whenever a message is sent to a host, an acknowledgement packet is returned. Periodically, the packets may be exchanged just to make sure the connection has not been lost. Each host will notify the other when the connection is to be closed.

2) UDP (User Datagram Protocol)—This protocol provides connectionless, unreliable delivery services. It assumes that the upper layer protocol will take the responsibility for ensuring that packets are acknowledged. It lets the upper-level applications send sections of messages, or datagrams, without extra overhead. UDP does not acknowledge the packets, request a virtual connection, or check to make sure that the data was received. You may use UDP to broadcast messages to all other hosts on the network. If a host needs to be notified that its message reached its destination, the application will acknowledge it. Upper-level protocols that use UDP protocol include TFTP, NFS, and Broadcasts.

UDP is like sending a post card in the mail. To send a post card, you do not need to contact the other party first. You simply write your message, address it, and mail it. This is similar to UDP. To send data to a remote host, the data is simply transmitted. No acknowledgement is expected. The UDP protocol is faster than TCP at the expense of ensuring reliable delivery.

Layer 2: Internet Protocols

This layer is responsible for routing data between hosts. It is also responsible for finding the "best" router. Best router sometimes means the fastest. Internet layer protocols are responsible for correctly routing packets across the internetwork. Connecting different local area networks together may create a complicated maze. Therefore, the hosts may be located on different networks separated by several routers. Here is a list of the Internet layer protocols.

1) IP (Internet Protocol)—This protocol is primarily responsible for the addressing of computers and the fragmentation of packets. It provides datagram service between communicating hosts. It also performs routing, fragmentation, and reassembly of datagrams.

2) ARP (Address Resolution Protocol)—This protocol translates a software address provided by IP into a hardware (MAC) address that is used by the Network Access layer. ARP functions are as follows:

    a) IP presents a datagram to the Network Access layer, which searches a temporary table for a physical address to match up with the IP address for the destination host. If an entry exists, the packet is sent to that physical address.

    b) If no entry exists, an ARP broadcast is sent out on the network requesting the physical address of the intended host.

    c) All PCs on the network receive the ARP broadcast and determine whether the requested address is the same as its own. The host with the same IP address as was requested replies back to the originator with its physical address.

    d) The originator host updates its table and sends the initial datagram to the receiver with the proper physical address.

3) RARP (Reverse Address Resolution Protocol)—This protocol functions in the reverse order as ARP. It translates a hardware address to a software address. RARP is used when a diskless workstation, which already knows its own hardware address, needs to acquire an IP address.

4) BootP—This protocol is used by diskless workstations to discover the IP address, discover the address of the server host, and discover the name of a file that is to be loaded into memory and executed at boot up.

5) ICMP (Internet Control Message Protocol)—This protocol is responsible for transporting error and diagnostic information for IP. It lets routers or hosts send error or control messages to other routers or hosts. ICMP allows a router with a full buffer to tell other routers to use another route. It allows a host to determine if another host is reachable.

Layer 1: Network Access Protocols

This layer is responsible for the physical connection between hosts. It defines the specifications related to the physical, or hardware, medium for data transmission, such as network interface boards, cabling, and network topology. The main functions of this layer are as follows:

a) Receiving an IP datagram and framing it into a stream of bits for physical transmission.

b) Specifying the hardware (MAC) address.

c) Making sure that the stream of bits making up the frame has been accurately received.

d) Specifying the access methods to the physical network, such as contention-based for Ethernet and token-passing for Token ring.

e) Specifying the physical media, the connectors, electrical signaling, etc.

The IEEE Project 802 established standards that define interface and protocol specifications for various network topologies. For LAN-oriented protocols, you have Ethernet (10Base5, 10Base2, and 10BaseT), Token Ring, and ARCnet. For WAN-oriented protocols, you have PPP (Point to Point Protocol), X.25, and frame relay. For example, on an IntranetWare server, you may bind TCP/IP to the LAN and WAN ODI (Open Data-Link interface) drivers.

As data is passed down from a client application, it must pass down through the four layers of the DOD model on the local system. It is then passed up the protocol suite on the remote host to reach the daemon process. Each layer adds a header and processes the data for the lower layer as shown below:

- Process/Application layer—An application, such as SMTP, passes its data to the Host-to-Host layer.

- Host-to-Host layer—TCP or UDP adds a TCP header to the data unit. The TCP header contains the source and destination ports identifying the upper layer protocol. It contains sequence, acknowledgement numbers, and header size, flags that establish, control, or terminate the connection. The TCP header also contains the maximum amount of data that the host is willing to accept and it uses checksum to guarantee data integrity. The message is then passed down to the Internet layer.

- Internet layer—The Internet layer adds an IP header to the message. The IP header contains information used to route the packet across the internetwork to the remote host. The IP header includes information, such as the software addresses of the source and destination hosts, the Host-to-Host (TCP or UDP) that is to receive the message, and flags controlling the fragmentation and reassembly of the packet. In addition, it contains segmented packets with an ID number, size of the packet, a header checksum, and TTL (Time-To-Live) to limit the life of the packet. The packet is then passed to the network access layer.

- Network Access layer—The network access layer adds a MAC (Media Access Control) header to the packet. On Ethernet networks, a frame includes information such as the MAC header containing the source and destination hardware addresses, IP header, and frame sequence containing the CRC (Cyclic Redundancy Check) of the MAC header. CRC is designed to ensure data integrity.

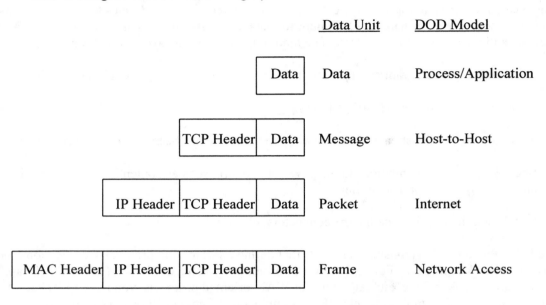

Figure 7.1 Passing data through the DOD model protocol stack

The four layers of the DOD model reside on the client and the server systems. Each TCP/IP protocol on the client side communicates directly with a corresponding TCP/IP protocol on the server side. When the frame reaches its destination, data passes up through the same four layers. Each layer strips off the appropriate header, processes the data, and passes the remaining data to the next layer until it reaches the application.

Surfing the Internet

There are several methods for finding documents on the web. You may browse home pages or hot lists of popular documents. You may also use hierarchical web directories, automated web search engines, or hybrid tools for a combination of web directories and search engines. There are three main components of automated web search tools. They are:

1) Spider—It is sometimes referred to as Robot. It is a program that searches the web every day, and is recorded in a catalog of spiders.

2) Catalog—It is a database of information about web documents that have been visited by spiders.

3) Search Engine—It is a program that takes a user query, matches it against the catalog, and displays information about the "relevant" documents to the user.

One must have general search strategies before using a search engine. You must understand the characteristics of each search tool and learn and use the search language of the search tool. Hierarchal web directories and hybrid tools are best for searching the "top" of the web. Within Yahoo, for example, you may select the proper category before doing the search. Automated search engines such as Yahoo, Lycos, InfoSeek, Alta Vista, and WebCrawler are best for finding documents on unusual topics. Make sure to use discriminating terms in your query to reduce the number of matches, and avoid one-word queries. Here are several popular automated web search engines.

- Lycos—It is operated by Lycos, Inc. and was developed by Dr. Michael L. Mauldin, Center for Machine Translation, and Carnegie Mellon University. It is one of the two largest catalogs on the Internet. The catalog covers "over 90%" of the web. It has indexed over 18,000,000 documents and is maintained by Robots. It has been generally ranked second (behind InfoSeek) in several studies. The information about each document is stored using the following:

    1) URL.
    3) Title.
    4) Outline (the first 200 characters in headings and subheadings.)
    5) Keys (100 most "weighty" words.)
    6) Excerpt (the smaller of the first 20 lines or 20 % of the document.)
    7) Date the document was last loaded.
    8) Date the document was last modified.
    9) Size of the document in bytes.
    10) Number of words.
    11) Description (up to 16 lines of "hyperlink" text from other documents to this document.)

Figure 7.2 Lycos web site

The information about matching documents is displayed according to a relevance score. Relevance is based on how many external documents point to the given document. Lycos adds, deletes, and updates about 50,000 documents a day in its catalog. The catalog is going faster than the Internet itself. URLs submitted by users are guaranteed to be indexed within one week. You must put a period after any search term that should be treated as word and not as a substring. You may use the detailed search form to control the number of search terms that must be in the document, specify the desired "closeness" of the match, or specify display options.

- Yahoo—This is an acronym for Yet Another Hierarchical Officious Oracle. David Filo and Jerry Yang developed it when they were Ph.D. candidates at Stanford University in April 1994. It is a privately funded company that provides the Yahoo services. Humans and not computers maintain it, and users submit it links. A Robot also looks for new site announcements at various places. It has indexed over 100,000 documents. The two ways of finding information are by browsing through subject categories and by searching keywords globally or within a selected category.

The information about each document is stored using the document title, URL, or comments supplied by user. Search results are displayed alphabetically by the category they were found in. The search results may consist of:

1) Names of documents that match the keywords.
12) Names of subject categories that matching documents belong to.
13) Tags that identify the best documents.
14) Names of subject categories that match the keywords.
15) The [Xtra!] tag that leads to the Reuters news-feed for that subject.
16) Some tags that indicate that the document was added in the last three days.
17) The "@" tag at the end of a subject category that indicates that this category appears at multiple places. Clicking on the heading will send you to the main category.

Figure 7.3 Yahoo.com web site

You may perform the following Yahoo search options. Remember, the search options are available on the detailed search screen only:

1) Make the search case-sensitive or case-insensitive.
2) Specify whether you want matches to contain all of your keywords or at least one of your keywords.
3) Specify whether the keywords should be considered as substrings or whole words.
4) Limit the number of matches found.

- InfoSeek—Robots maintain it although some pages have been reviewed by humans. InfoSeek has been rated as the best search engine in several tests. It has a relatively sophisticated search language and does not support Boolean search operators. It offers a free-based searching of the web and other databases. Search results are sorted by relevance and displayed 10 at a time. There are three factors that affect the relevance score for a document:

1) The number of times the keywords appear in the document.
2) More weight is given to the existence of more discriminating terms. For example, "Clinton" is likely to generate a higher relevance score than "president".
3) Existence of phrases being searched for generates a higher relevance score.

InfoSeek has a relatively powerful search language. Make sure to use discriminating keywords and do not expect to find documents that appear in the lower levels of the web. You may do the following when using the InfoSeek search engine:

1) Use capital letters to indicate proper names.
2) Put double quotes around words to indicate that the words must appear next to each other.
3) Place a hyphen between words to indicate that the words must be within one word of each other.
4) Put square brackets around words to indicate that the words must be within 100 words of each other.
5) Use the plus sign "+" between words to indicate that this word must appear in this document.

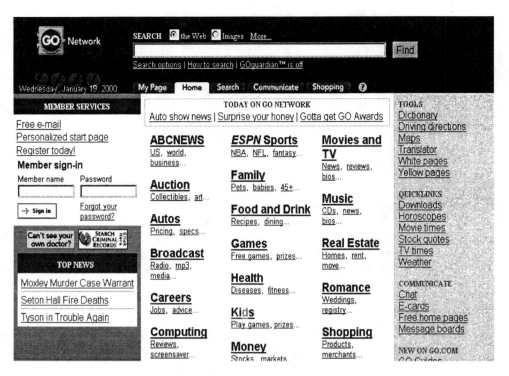

Figure 7.4 InfoSeek.Go web site

- Alta Vista—Developed by Digital Equipment Corporation (DEC—now owned by Compaq Corp.), it is the newest automated search engine. It gives access to over 8 billion words found in over 16 million web pages. You may perform the following tasks using the Alta Vista search engine:

  1) Search for phrases by putting them in quotes.
  2) You may specify required keywords, prohibited keywords, and wildcards. For example: +president* - "foreign policy".
  3) You may find out what has links to a given URL. For example, to find out which external documents have links to the America Online home web page : +Llink:http://www.aol.com –url:http://www.aol-shopping.com.
  4) You may search for words in titles of documents only. For example, to find all documents that have "New Jersey" in the title tag: title:"New Jersey".

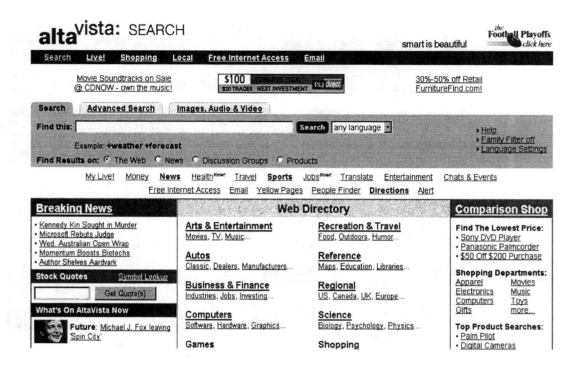

Figure 7.5 Alta Vista web site

- WebCrawler—Operated by America Online, it was developed by Brian Pinkerton at the University of Washington. It is maintained by Robots and has indexed over 250,000 documents. Information is stored for each document according to document title or keywords. Titles of matching documents are displayed according to a relevance score. Relevance score is calculated by taking the total number of times each of the words in your query appears in the document and dividing it by the total number of words in the document. You may see more hits on one page because only titles are displayed. WebCrawler has an unlimited number of hits, does not have a rich search language, and uses stop words in indexing.

Keep in mind, the web is growing very fast. Human generated catalogs may not be able to keep up with the growth. Information in catalogs may become obsolete. The search space is not clearly defined and is constantly changing. Document set is highly heterogeneous with respect to type, content, and style. User interface designs for search engines are always changing. Finally, there are no standards for Robot exclusion.

Figure 7.6 WebCrawler web site

Creating Web Pages Using HTML

HTML (Hyper Text Markup Language) is commands embedded in a document. When an HTML document is sent to another computer, HTML commands tell a web browser program how to display the text and graphics within the page. HTML links are highlighted words or pictures that have "click points" that request another HTML document. The function of HTML is to manage text and the way the text is viewed. It also provides links to internal and external documents.

HTML connection happens when a client sends a "Connect Request" to a web server and the web server gives "Permission To Talk" back to a web browser. When a client sends a "Document Request" to the HTML server, the server responds with an HTML document to the client web browser. Each machine sends a "Disconnect" to the other and the cycle ends. The web browser's function is to convert HTML codes to GUI when receiving HTML web pages. The browsers convert GUI to HTML codes when requesting documents. Examples of web browsers are: Netscape, Microsoft Internet Explorer, Mosaic, etc.

A URL is used to provide a resource name, Internet address, path, and filename of the HTML document. For example:

URL = http://www.aol.com/index.htm
Resource name = http://
Internet address = www.aol.com
Path and filename = index.htm

Follow the following steps when planning your web page:

1) Define the audience and ask what they want.
2) Decide on the purpose of the web page.
3) Set specific goals and objectives.
4) Decide on what information you want to provide.
5) Decide on how to arrange the information and the order of presentation.
6) Decide on how to break it up into digestible chunks.

Follow the following steps when designing your web page:

1) Apply all the planning criteria, decide on the physical layout of the web page, and keep it simple.
2) Design the web page with tree structures and for all types of web browsers.
3) Make your page fast loading. Remember that graphics will always make slower page loading. Keep graphics small and worthwhile. Only use graphics to explain a complex topic.
4) Use a consistent graphic look. Make sure to have a logo on each page and to title all the pages. Make sure to have the links and choices in the middle and link-backs at the bottom of the page.
5) Make organization obvious. Have home page links to detailed pages.

Follow the following steps creating your web page:

1) Create an HTML document from scratch by:

   a) Opening a word processor or text editor.
   b) Writing the text of your document. Save the file as an ASCII or DOS file.
   c) Placing the HTML tags into the document.
   d) Saving the document with the .htm or .html extension.

2) View the HTML document by:

   a) Exiting the word processor.
   b) Starting the web browser.
   c) Choosing the option to view a local file.

3) Remember that tabs, spaces, and returns are ignored within the document text.
4) Remember that upper and lower case do not matter inside HTML tags.
5) You may place tags in any column.
6) Be brief, and spell check and proofread all documents.
7) Use headings to separate and summarize.
8) Use lists for important information.
9) Make it easy to scan and make each page complete in itself.
10) Provide links between pages and use short phrases as links.
11) Keep heading sizes constant and give generic instructions.
12) Use bold, italics and caps sparingly and limit the size and number of images.
13) Use alternative image descriptions and sign and date your document.

HTML Tags

HTML tags are information for browsers to describe how a document is structured. HTML tags do not affect how a document looks. They are enclosed in <> angled brackets. The tag name is inserted inside the brackets (e.g., <title>). Tags mostly come in pairs but they are single commands (<bold>....</bold>). The start tag <bold> tells the browser when to start interpreting the text as bold. The stop tag </bold> tells the browser when to stop interpreting the text as bold. Remember that HTML tags are case insensitive (<BOLD> = <bold>). Table 7.1 shows a list of the basic HTML tags.

Tag Name	Tag Symbol	Example	Comment/Description
HTML	Pair <html></html>	<html> ….. Document </html>	They are the first and last tags you write.
Head	Pair <head></head>	<html> <head> <title> …. ACME Engineering </title> </head> …. Document </html>	Contains only a <title> tag. Do not put any other text in the tag.
Title	Pair <title></title>	<title> Acme Engineering </title>	They are placed inside the <head></head> tags. The document must have one unique title per page. The title must only contain text. The title is displayed in the title bar of the browser and bookmark list of browser.
Body	Pair <body></body>	<html> <head> <title> Acme Engineering </title></head> <body> … document text </body> </html>	These indicate the beginning and end of text body.

Table 7.1 Basic HTML tags

Table 7.2 shows a list of the structure tags. Structure tags create the basic look of a page. The structure displays the information.

Tag Name	Tag Symbol	Example	Comment/Description
Heading or Header	Pair <h1></h1>	<h1> Acme Engineering </h1>	They are used to separate sections of text. There are six kinds of headings (h1, h2, h3, h4, h5, h6). The higher the value, the smaller the text font size. The text of headings can be of any length. Use the level 1 heading tag at the start of a document. Use numerical order and do not skip numbers. As a suggestion: use only h1, h2, and h3. h4 and above are too small.
Paragraph	Pair <p></p> or single <p>	<p> This is text </p> <p> This is text </p> This is text This is text	The pair of tags places space between paragraphs of text. The single tag places a space between blocks of text. Do not use the single tag to create white space in the document.
Horizontal Line or Horizontal Ruler	Single <hr>		Draws a horizontal line across the page. Do not include a line in the middle of text. Use it to separate sections of documents. Use it sparingly.
Text Break	Single 		Do not include any text, just the tag. Use it at the end of a line to restart text on the next line. It is the same as the carriage return when typing.
Comment	<!—your comment here .. >		Does not display in browser. It appears only in the HTML source document.

Table 7.2 Structure tags

Tables 7.3 and 7.4 show a list of the appearance tags. There are two types of appearance tags: logical and physical. Logical tags indicate how text is being used. Physical tags indicate how a local browser is displaying text. Not all browsers can handle physical tags.

Tag Name	Tag Symbol	Example	Comment/Description
Emphasized Text	Pair <em></em>	<em> Good </em> morning class. *Good* morning class.	For graphical browsers, it makes the text italics.
Strongly Emphasized Text	Pair <strong></strong>	Good morning <strong> class </strong>. Good morning **class**.	For graphical browsers, it makes the text bold.
Programming Text	Pair <code></code>	<code> Good morning class </code>. `Good morning class.`	For graphical browsers, it makes the text Courier font.
Sample Text	Pair <samp></samp>	<samp> Good morning class </samp>. `Good morning class.`	For graphical browsers, it makes the text Courier font.
Keyboard Text	Pair <kbd></kbd>	<kbd> Good morning class </kbd>. `Good morning class.`	For graphical browsers, it makes the text Courier font.
Variable Text	Pair <var></var>	<var> Good morning class </var>. *Good morning class.*	For graphical browsers, it makes the text italics and underlined.
Definition	Pair <dfn></dfn>	<dfn> Good morning class </dfn>. **Good morning class.**	For graphical browsers, it makes the text bold.
Citation	Pair <cite></cite>	<cite> Good morning class </cite>. *Good morning class.*	For graphical browsers, it makes the text italics.

Table 7.3 Logical appearance tags

Tag Name	Tag Symbol	Example	Comment/Description
Bold	Pair \<b>\</b>	\<b> Good \</b> morning class. **Good** morning class.	For graphical browsers, it makes the text bold.
Italic	Pair \<i>\</i>	Good morning \<i> class \</i>. Good morning *class*.	For graphical browsers, it makes the text italics.
Typed Text	Pair \<tt>\</tt>	\<tt> Good morning class \</tt>. `Good morning class.`	For graphical browsers, it makes the text Courier font.
Nested physical	Two Pair \<b>\<i>\</i>\</b>	\<b>\<i> Good morning class \</i>\</b>. **Good morning class.**	You may nest tags together. For graphical browsers, it makes the text bold and italics.

Table 7.4 Physical appearance tags

Tag Name	Tag Symbol	Example	Comment/Description
Numbered	Pair \<ol>\</ol>, pair \<lh>\</lh>, or single \<li>	\<lh> list of headings \</lh>	ol stands for ordered list. The browsers will display this as a numbered list. If the list is edited, it will automatically be renumbered.
Bulleted	Pair \<ul>\</ul>, pair \<lh>\</lh>, or single \<li>	\<lh> list of headings \</lh>	ul stands for unordered list. The browsers will display bullets or asterisks before each list item. Use when order does not matter.
Definition	Pair \<dl>\</dl>, single \<dt>, or single \<dd>	\<dl> list of headings \</dl>	dt stands for definition term. dd stands for definition definition. The browsers will display it as an indented definition. Use any time you need an indented list.
Menu and Directory	Pair \<dir>\</dir>, pair \<menu>\</menu>, or single \<li>	\<menu> list of headings \</menu>	The simplest form of list and bulleted text.

Table 7.5 List tags

Tag Name	Tag Symbol	Example	Comment/Description
Preformatted Text	Pair <pre></pre>	<pre>.. all imported text …</pre>	It will keep all the spaces you have and always displays Courier font. You may use it to create columns, keep 70 line characters or less, and have quick conversions of imported text.
Address	Pair <address></address>	<address> Mike's Home page copyright 2000 </br></address>	It is used for information about the user's page. Your signature on the page, which is at the bottom of the page, may include the date, a copyright symbol, or your e-mail address.
Anchor	Pair <a></a>		It contains name of tag and tag attributes. Attributes define what kind of link it is. The most common attribute is HREF (Hypertext Reference).
Image	Single <img>		Used for image links. To be discussed later in chapter.

Table 7.6 Miscellaneous tags

Remember that you may combine list types together. You may place one type of list inside another. This feature is good for outlines and tables of contents. Browsers will display lists differently. Indent tags in an HTML document make the document have an easy-to-see structure. In addition to tags, you have some special characters. The special characters are equivalents for characters not on basic keyboards. There are two kinds of entries: named and numbered. For example, the HTML characters <, >, " and & need equivalents. Table 7.7 shows a complete list of the special characters codes:

Character	Named	Numbered
<	&lt	&#60
>	&gt	&#62
"	&quot	&#34
&	&amp	&#38
(		&#40
)		&#41

Table 7.7 Special characters codes

The following are some examples on how to use the special characters:

"This is &quot very &quote important" or "This is &#34 very &#34 important"

HTML Links

Links are used to specify file names and text click points to retrieve local or remote documents. Links can be text, graphics, sound, video, or any other OLE. Here are steps to create a link to a local document called techinst.htm:

1) Place the anchor links around the text you want to use as a link.

   <a> Other Technical Institutes </a>

2) Add the name of the file you want.

   <a "techinst.htm"> Other Technical Institutes </a>

3) Add the attribute before the file name.

   <a HREF="techinst.htm"> Other Technical Institutes </a>

Links may also be pointed to HTML documents on your computer or other resources on the Internet. For example:

<a HREF="resume.htm"> My Resume </a> or
<a HREF="http://www.aol.com"> America Online home page </a>

When incorporating links into text, be sure to write the document first. Identify one or two words for a link. Be descriptive (sensitive to flow). Use the links to link menus to organize lots of links and avoid "click here" links. For example, you may create "Mailto" links to automatically be linked to send an e-mail to somebody on the Internet. Remember that the "Mailto" links are part of the <address> tag. It contains a link to an email form. For example:

<a HREF="mailto:mawwad@admin.nj.devry.edu"> mawwad@admin.nj.devry.edu </a></address>

You may cut and paste a URL to reduce errors in copying URL names:

1) Locate the URL in the text document.
2) Highlight the URL with a mouse drag.
3) Edit copy (Alt-E-C) or (Ctrl-C).
4) Switch to the HTML document.
5) Edit paste (Alt-E-P) or (Ctrl-V).

In addition to text links, you may use image links to link to other HTML documents. Image links have two forms, internal and external: Internal links are inline images that are sent as part of the web page and reside on the web server. They are sent to the browser when the page is loaded. External image links are also with web pages, but they reside on someone else's computer. External images are accessed via a link. Beware of copyright infringements and limit image links to necessary images only. Make sure to do the following for all images:

1) Limit the size (< 20 KB) and number of colors.
2) For text descriptions, use the ALT= for images.
3) Test images in black and white and color.
4) Test the images on multiple browsers.

The single image tag <img> uses three attributes, SRC, ALT, and ALIGN. Table 7.8 shows, the three image attributes and how they are used:

Attribute	Example	Description
SRC	Single tag src= <img src = "devry.gif"> (internal link) <img src = "http://www.devry.edu/gifs/devry.gif"> (external link)	It indicates the name of the image to be displayed.
ALT	Single tag alt= <img src = "picture.gif" alt = "[picture of logo]"> <img src = "picture.gif" alt = " "> <! Display nothing>	This is for text-only browsers or if inline images are turned off. It can define what text people see and [IMAGE] is the default in text browsers.
ALIGN	Single tag align= <img src = "picture.gif" align = middle | top | bottom> My picture is here.	It controls how the text and images are related to each other. It can align to top, middle, or bottom. It only affects the current line. It tells text where to position itself.

Table 7.8 Image tags attributes

You may use an image as a link with text descriptions, for example:

<a HREF = "http:// www.devrycols.edu">< IMG SRC = "devry.gif"> Back to Home Page </a>

Images can go anywhere in the text. If the browser cannot display the file type, it finds a viewer. Indicate the format and size of the file, inside or following the link. For example:

<a HREF = "picture.gif"> My picture (100K GIF file)</a> or
<a HREF = "picture.gif"> My picture </a> (100K GIF file)

Make sure to thumbnail image links; limit the size of the photo, so that the page loads fast. Make sure you view the picture before downloading it. Create a small version of the picture and place it in a directory along with large versions. For example:

<a HREF = "bigpict"> <IMG SRC = "littpict.gif"> </a>

You may also use sound and video links. You must be connected via an internal or an external link. If the browser cannot display the video or play the sound, it finds a viewer or a player and asks the user to save or play or run the sound or video. Make sure to indicate the format and size of file in the links, for example:

<a HREF = "song.au"> This is our song </a> (AU Format 400K) or
<a HREF = "swim.avi"> This is a swimming video <AVI Format 1600K) </a>

Create the following home page:

<HTML>
<HEAD><TITLE> ACME Engineering Home Page </TITLE></HEAD>
<BODY><CENTER>
        <h1> Welcome to the ACME Engineering Home Page </h1>
        <h2> 1234 Central Ave
            Wayne, New Jersey 07470
            Tel 973.123.4567 </h2></CENTER>
Write anything that you want in this part of the page and say that if you have any questions, e-mail me at: (Include your Mailto address here)   </BODY>
</HTML>

IP Addressing

One of the most important topics in any discussion of TCP/IP is IP addressing. An IP address is a numeric identifier assigned to each workstation on an IP network. This is a software address and not a hardware address that designates where the workstation resides. The most important duty of the network administrator on an IP network is to assign and maintain IP addresses for the hosts (workstations, or nodes) on the network.

IP addressing is the most complex part of TCP/IP. For example, an Ethernet hardware (MAC) address is 48 bits. An IP address is 32 bits. As a result there is no direct match between the two numbers. That is why ARP is needed to match the numbers in a cached table. The computer needs the MAC address and TCP/IP needs the IP address. It is not easy to remember a 48-bit MAC address or even a 32-bit IP address. By convention, the IP address is usually represented in a dotted decimal notation, such as the IP address 204.78.135.15. Each segment of the network on the internetwork is also assigned a network address, which is represented by a portion of the IP address of each host on the network. From the user's point of view, it is convenient to associate a name to each host or network.

To make it practical, so that you do not have to remember IP addresses, the following three methods are commonly used to map hostnames (ASCII names that you can recognize) to IP addresses.

1) Host Tables—A host table is a text file that maps the commonly used host name to that host's IP address. It translates the host name of a machine to its IP address, or software address. On a UNIX system, the host file hosts is located in the /etc directory. On the IntranetWare server, the file HOSTS will be in the directory SYS: ETC\. Below is an example of a hosts table:

# Internet Address	Host Name	Aliases	Comment
127.0.0.1	localhost	lb	# loopback
192.67.67.20	NIC.DDN.MIL		# DOD domain
255.255.255.255	broadcast		

For each host, one line should be represented in the host file with the format that each host file entry provides information about one host on the Internet. In addition, any number of blanks and/or TAB characters separate items. The format of the host table of SYS: ETC\HOSTS is equivalent to the host table in UNIX systems /etc/hosts. From the above host table the format is as follows:

[IP Address]   [Host Name]   [Alias] [Comments]

a) IP Address—The internet address is a 4-byte address in standard dotted decimal notation. Each byte is a decimal or hexadecimal value separated by a period. Hexadecimal numbers must start with the character pair "0x" or "0X". This address is assigned to the host. On the NetWare 4.11 server, this assignment is made when you bind the protocol to the network board. On the workstation, the IP address is assigned in the NET.CFG file.

b) Host Name—The host name is the name of the system with the given address. Each host name should be unique and can contain any printable character other than a tab, space, new line, or a number sign "#". This name is arbitrary and is chosen by the system administrator. The host name for a NetWare server should be the same as the server name.

c) Alias—The alias is another name for the host. A host may have many aliases. A space is required between each name.

d) Comments—A "#" indicates the beginning of a comment. You may place comments at the end of a host name or on a line by itself. Routines that search the file do not read characters following a "#".

Another file that is similar to the host table is the network table. A network table contains names and network addresses of known networks. On UNIX systems, the network file is the file etc/networks, and on NetWare 4.11 server, the host file will be SYS: ETC/NETWORKS. Below is an example of a network table:

# Network Name	Network Address	Alias	Comment
loopback	127		# loopback network number
Novell	130.57		# Novell's network number

For each network, there should be one line in the network file with the format that each network file entry provides information about one network on the Internet. In addition, any number of blanks and/or TAB characters may separate items. From the above network table the format is as follows:

[Network Name]   [Network Address]   [Alias] [Comments]

a) Network Name—The network name is the network name assigned to this network number. Each host name should be unique and can contain any printable character other than a tab, space, new line, or a number sign "#".

b) Network Address—The network address is the number of the network. It is the network portion of the IP address for all hosts on that network. Hexadecimal numbers must start with the character pair "0x" or "0X".

c) Alias—The alias is another name for the network. A network may have many aliases. A space is required between each name.

d) Comments—A "#" indicates the beginning of a comment.

2) DNS (Domain Name System)—DNS is widely used on the Internet to translate host names to IP addresses. It is a mechanism that helps users to locate the name of a machine and to map a name to an IP address. Because the Internet contains thousands of systems, it is possible for two different computers to have the same host name. To avoid confusion, the Internet has a number of domains. Machines throughout the Internet called name servers keep a database of large numbers of host names. The database is arranged in a hierarchical manner, starting with the root and moving down to the domain, sub domain, and finally to the host name. The Domain Name System hierarchy is shown in Figure 7.7.

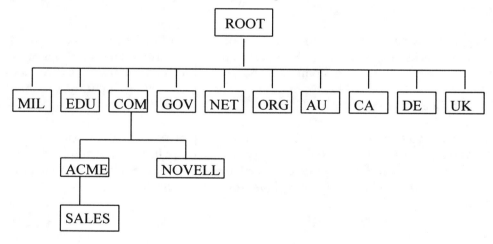

Figure 7.7 Domain Name System (DNS) hierarchy

From Figure 7.7:

Domain	Acronym	Description
MIL	Military	Used by the Department of Defense.
EDU	Education	Used by colleges and universities.
COM	Commercial	Used by corporations or businesses.
GOV	Government	Used by government organizations.
NET	Network	Used by an administrative organization for a gateway or a network.
ORG	Organization	Used by other organizations.
AU	Australia	
CA	Canada	
DE	Germany	
UK	United Kingdom	

DNS is structured as an inverted tree, much like the directory structure on an IntranetWare volume. Each node on the tree represents a domain. Each domain has subdomains, which can have further divisions as necessary. Name servers are set up to maintain host addresses for each subdomain or zone within the name space as well as the addresses of the root name servers. By using recursive queries or referring clients to other servers, a DNS server is capable of resolving the IP address for any host on the Internet. The domain name for a host consists of a set of domains separated by periods. The format is HostName.Subdomain.Domain. In Figure 7.7, you have the domain SALES.ACME.COM.

3) NIS (Network Information Services)—NIS is another type of name services commonly used on IP internetworks. You may use NIS servers to provide host-to-address translation within domains. Domains are groups of computers. NIS servers contain databases, called maps. In addition to host names and addresses, maps may contain user and group information. Maps are shared by a group of computers.

When you connect two systems together using TCP/IP, each protocol layer on one host communicates directly with the corresponding layer on the other host. For example, the network access layer on one host communicates directly with the network access layer on another host. Each DOD layer identifies a host using one of the following addressing schemes.

DOD Layer	Addressing Method
Process/Application	Host name
Internet	IP Address
Network Access	Hardware address

An IP address is made up of 32 bits of information. These bits are divided into four bytes (octets). There are two methods of representing an IP address:

1) Dotted-decimal, as in 204.65.156.25 or
2) Binary, as 11001100.01000001.10011100.00011001

Both of these examples represent the same IP address. Each TCP/IP host must have an IP address. All hosts must have a software address at the Internet layer. The IP address (Internet address) is a software address for a host in a TCP/IP network. The IP address identifies the network to which the host is attached. Routers use the IP address to forward messages to the correct destination. For each host the IP address is divided into two parts—a network address and a node (host) address. The network portion of the IP address must match the network address of every other host on the same network. The host (node) portion of the IP address must be unique.

For example, an IP network has been assigned the network address of 140.56. Its network address uniquely distinguishes it from all other networks on the internetwork. Each host or node on this network must have an IP address of the form 140.56.x.x where the last two octets of the address must be unique on the network. A node assigned the last two octets of 25.150 is uniquely identified on the internetwork by using 140.56.25.150 as its address.

There are five classes of IP addresses. The chart below sets the groundwork for understanding IP addresses:

Class	First byte address range	Leftmost bit pattern	Network/host designation
A	0 to 127	0	N H H H
B	128 to 191	10	N N H H
C	192 to 223	110	N N N H
D	224 to 239	1110	Multicast addresses
E	240 to 255	1111	Internet experimentation

From the above chart, column A shows the class type. Column two shows the number range for the first byte that defines the range. Column three shows the binary bit pattern that is the beginning of the range byte. This pattern is helpful if you forget which class an address belongs to. Column four show which bytes are for a network (N) address and which are for a host (H) address. Let us describe each class in a little more detail:

- Class A—The first byte is the network address. The first bit must be zero. The next three bytes are for the node (host) address. On the Internet, all class A addresses are already assigned. There are 127 class A networks. Each class A network may have up to $2^{24} - 2 = 16,777,214$ hosts per network.

- Class B—The first two bytes identify the network address. There are $2^{14} = 16,384$ class B networks and $2^{16} - 2 = 65,534$ hosts per network.

- Class C—The first three bytes define the network address. There are $2^{21} = 2,097,152$ class C networks and 254 hosts per network.

- Class D—The range is from 224.0.0.0 to 239.255.255.255.The addresses are used for multicast packets. Multicast packets are used by many protocols to reach a group of hosts. ICMP router discovery is a protocol that uses multicast packets. A host can determine the addresses of routers on its segment by transmitting a packet addressed to 224.0.0.2, which is received by all routers on the network.

- Class E—The range is from 240.0.0.0 to 255.255.255.255. Class E addresses are reserved for future addressing modes.

Class D and E addresses are not assigned to individual hosts on the internetwork. In addition, there are some reserved IP addresses for special purposes as shown below:

Address	Description
Network address of all 0s	Refers to "this network."
Node address of all 0s	Refers to "this node."
Network address of all 1s	Refers to "all networks."
Node address of all 1s	Refers to "all nodes" on a specified network.

When the entire IP address is set to 0, it refers to the default route. This route is used to simplify routing tables used by RIP protocols. When the entire IP address is set to 1, it refers to a broadcast to all nodes on the network.

127.0.0.0	Reserved for loopback. The address 127.0.0.1 often is used to denote or refer to the local host. Using this address, applications can address a host as if it were a remote host without relying on any configuration information.

NIC (Network Information Center) is responsible for assigning network addresses to all the computers on the Internet. For further information contact:

> Network Solutions
> InterNIC Registration Services
> 505 Huntmar Park Drive
> Herndon, VA 22070

You may also obtain help by e-mail at hostmaster@internic.net. Once the network address is assigned, you may assign addresses to individual hosts on the network by using the same network address and a unique host address for each node. For example, if you have a class B network address of 140.67.0.0, you may assign any of the following addresses to the hosts within the network:

140.67.59.32
140.67.0.19
140.67.125.234

Questions

1) List all the popular services available on the Internet.
2) List all the layers of the OSI model, and what the data is called at each layer.
3) Describe the function of each layer of the OSI mode.
4) Where do the following devices operate in the OSI model? Describe the function of each.

   Gateway, Repeater, Bridge, Router

5) List all the layers of the DOD model.

6) How does the DOD model compare to the OSI model?

7) List all the protocols in each layer of the DOD model.

8) Write down what each acronym of the protocols stands for and describe what each one is used for.

9) Given an IP address = 209.35.67.35

    a) What is the host address? _____

    b) What is the network address? _____

    c) What is the class? _____

10) How many network and host (node) addresses can classes A, B, and C have?

11) What is the DOD layer that routes packets between different hosts or networks?

12) Name three protocols that provide file transfer functions.

13) What TCP/IP protocol provides terminal emulation?

14) What TCP/IP protocol uses the equivalent of the Transport layer of the OSI model?

15) What protocol provides connectionless delivery services of datagrams between hosts?

16) What protocol is responsible for guaranteed delivery?

17) What is the Internet layer protocol that translates IP addresses into MAC addresses?

18) What are the different tasks that may be performed in the search engines Yahoo, InfoSeek, Alta Vista, Lycos, and WebCrawler.

19) Explain the planning phase of a web page.

20) What is the difference between a physical and a logical appearance tag?

21) Name and describe the three attributes of the image tags.

22) Create your own home page. Make sure to link it to other documents using text and image links.

Index